LANDMARK COLLECTOR'S LIBRARY

# The Surviving Engines of Glyn Pits, Pontypool

## Early Steam and Water Power in Local Industry

Gwyn Tilley

Foreword by Sir Richard Hanbury Tenison, KCVO

Gwyn Tilley is uniquely qualified to write about the Glyn Pits, and has given us a fascinating record and some wonderful photographs of the present state of that important industrial site. For three hundred years the Pontypool ironworks, like other ironworks worldwide, had operated mainly in charcoal, but developments during the early 19th century opened up a new role for coal, and my great-grandfather was lucky to have access to easily won seams in the Glyn Valley. In 1850, however, after several years of loss, he decided to lease the various ironworks and their associate coal mines, and my family's long association with the industry came to an end.

Very little remains of the earlier ironworks and it is only the remoteness of the Glyn Pits that has saved them from general destruction. If it is ever possible to make the old pit buildings safe and secure, and so open to visitors (as I hope will be the case), Mr Tilley's record will form an essential tool.

Sir Richard Hanbury Tenison, KCVO

Published with the generous aid of
The Awards For All Wales, part of the
Heritage Lottery Fund

"I sell here, sir, what all the world desires to have –
POWER"

**Matthew Boulton, Engineer, 1728 – 1809**

Speaking to James Boswell of his engineering works

"All that mankind has done, thought, gained, or been;
it is lying, as if in magic preservation, in the pages of books"

**Thomas Carlyle, 1795 – 1881**

LANDMARK COLLECTOR'S LIBRARY

# THE SURVIVING ENGINES OF GLYN PITS, PONTYPOOL

## EARLY STEAM AND WATER POWER IN LOCAL INDUSTRY

GWYN TILLEY

Landmark Publishing

Published by

**Landmark Publishing Ltd**
Ashbourne Hall, Cokayne Ave, Ashbourne, Derbyshire DE6 1EJ England
Tel: (01335) 347349 Fax: (01335) 347303
e-mail: landmark@clara.net
website: www.landmarkpublishing.co.uk

ISBN 1 84306 203 8



British Library Cataloguing in Publication Data: a catalogue record for this book is available from the British Library.

**Print:** Bath Press Ltd, Bath

**Design:** Mark Titterton

**Cover:** James Allsopp

**Front cover and Title page:** Brake wheel, rope-wheels, main crank etc. on the vertical winder at Glyn Pits

**Back cover top:** A group of surface workers at Glyntillery Colliery, near Pontypool, in Gwent.
This mine, one of the wettest in the district, is about half a mile west of Glyn Pits.

**Back cover bottom:** Beam pumping engine.

# Author's Qualifications

The author, who is a Member of the Newcomen Society for the Study of the History of Engineering and Technology, has worked on steam equipment of most kinds nearly all of his working life. This ranged from feed and water pumps such as Wier and Worthington Simpson to steam/air compressors, such as the Bellis and Morcom steam/air compressors at Llanerch Colliery, etc. He worked on boilers of all types, including retubing locomotive and stationary engine boilers; general repairs to vertical types, as used in two local dairies (Clarence and Albion Road); horizontal types, such as Cornish boilers; which he worked on in the early 1940s at St Davids Hospital, Tredegar, and also Blaina gasworks, and which were the type originally fitted at Glyn Pit. Largest of all were the Lancashire boilers, used in most of the local deep mines, Blaenserchan, Llanerch, etc. and local tin and steelworks, and latterly Glyn Pit. He also worked on an 'Agricultural engine'. This small engine was designed to be horse-drawn between farms for corn thrashing, etc. and it was used latterly (1940s) at the Pontypool Council 'Destructor' on the Albion Road, where it drove a mortar mill. It is now incomplete and held at Llanyravon Farm Museum, Cwmbran. The author was the last person to retube the boiler of this engine in the 1940s.

Latterly, he worked on gas-fired steam generating boilers and heating equipment at the County Hospital, Griffithstown, and Pontypool and District Hospital, for Gwent Health Authority.

Another perhaps more subjective but nonetheless appropriate qualification is that he started work at the Blaendare Company's Quarry Level wagon repair shed near the Glyn Pit in 1944 before serving his apprenticeship with the Pontypool Welding Co. Ltd. During this time he attended evening classes at Abersychan Mining and Technical College for five years studying welding and metallurgy, before serving in HM Forces in the Royal Electrical and Mechanical Engineers (R.E.M.E.) and attaining the rank of Sergeant.

He had a regular weekly column for the *South Wales Argus* for six years, writing on a variety of subjects, including local history. He has known the Glyn Pit, and some of its personnel too, for a far longer time; since 1940, in fact, or nearly sixty-three years.

# ACKNOWLEDGEMENTS

When I decided to write this work on the Old Glyn Pits, near Pontypool, in Gwent, I had two purposes in mind: one, that I should do what I could to document its history and to help to preserve the Old Glyn Pit – and some other local sites – for future generations; and two, for the benefit of those who are not familiar with steam engines and mining technology, to explain, as well as I could, how the engines, condenser, ventilation, etc. worked.

In the first task, I have been greatly assisted by several people. First, Sir Richard Hanbury Tenison, KCVO, of Pontypool Park Estate (who once again owns the mine), was particularly helpful, allowing me to peruse the extensive collection of Hanbury papers held at the Park Estate Office, and who not only found the relevant papers for me, but even went so far as to place his own office at my disposal for their study. Also his son, Mr Jack Tenison and the staff at the Park Estate Offices, Pontymoile, for their unfailing courtesy and help over my many years' involvement with the Old Glyn.

I would also like to thank the following people: the staff of the County Record Office, Croesyceiliog, for their willing assistance; Peter Kennedy, who was instrumental in obtaining a lottery grant for me and Mr John Cornwell, of North Somerset, for both encouragement and also his kind permission to use his photograph of the vertical winding engine at the Glyn. The staff at Torfaen Museum Trust, including Mr Gordon Griffiths, photographer, for help in research and permission to use various photographs and maps; Mr David Hall, of Ponthir, for the use of some of his photographs of the Glyn, which, since all the top set of valve gear on the vertical winding engine is now missing, are virtually unobtainable; Councillor Neil Waite, for invaluable information on The Scourings at Upper Race, which is now a protected site, and the use of his photograph of tram repairers at Quarry Level and my friend and colleague, Mr Clive Davies, from both of whom I have learned much and last, but by no means least, my long-suffering wife Joan without whom I would have written nothing at all!

In the second task, explaining the machinery, I have relied on my own experience and knowledge of engineering. Therefore, any faults or omissions are mine.

Gwyn J. Tilley
Pontypool
August 2005

# CONTENTS

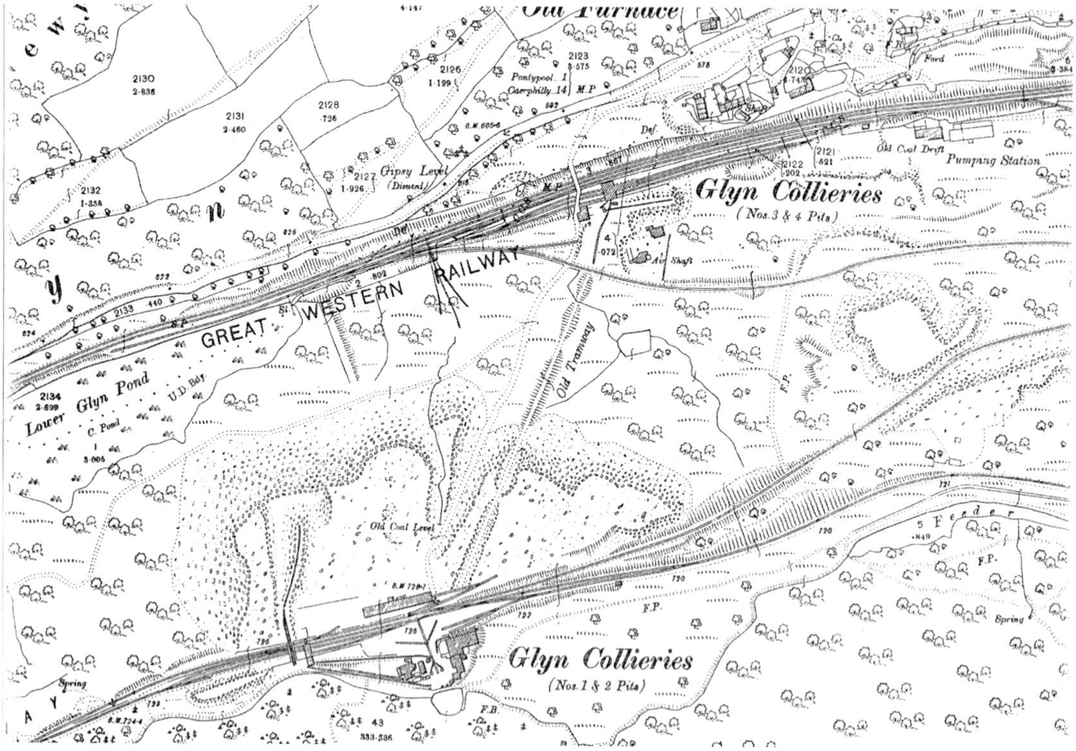


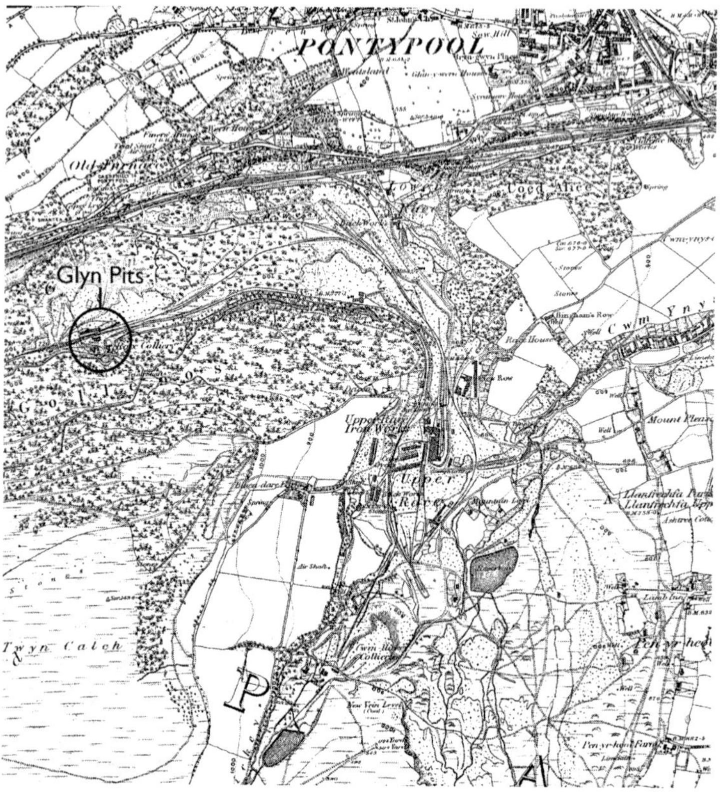


Fig 1: Map showing Old Glyn Pits – Glyn Collieries on map – in relation to Glyn Pits No. 3 and 4 pits, Park Estate line (mineral railway) and Old Furnace and main rail line to Crumlin, Caerphilly, etc. The mineral railway on bottom right was to Blaendare colliery, and later to the Balance Pits.

Fig 1a: Area Map showing Old Glyn Pits.

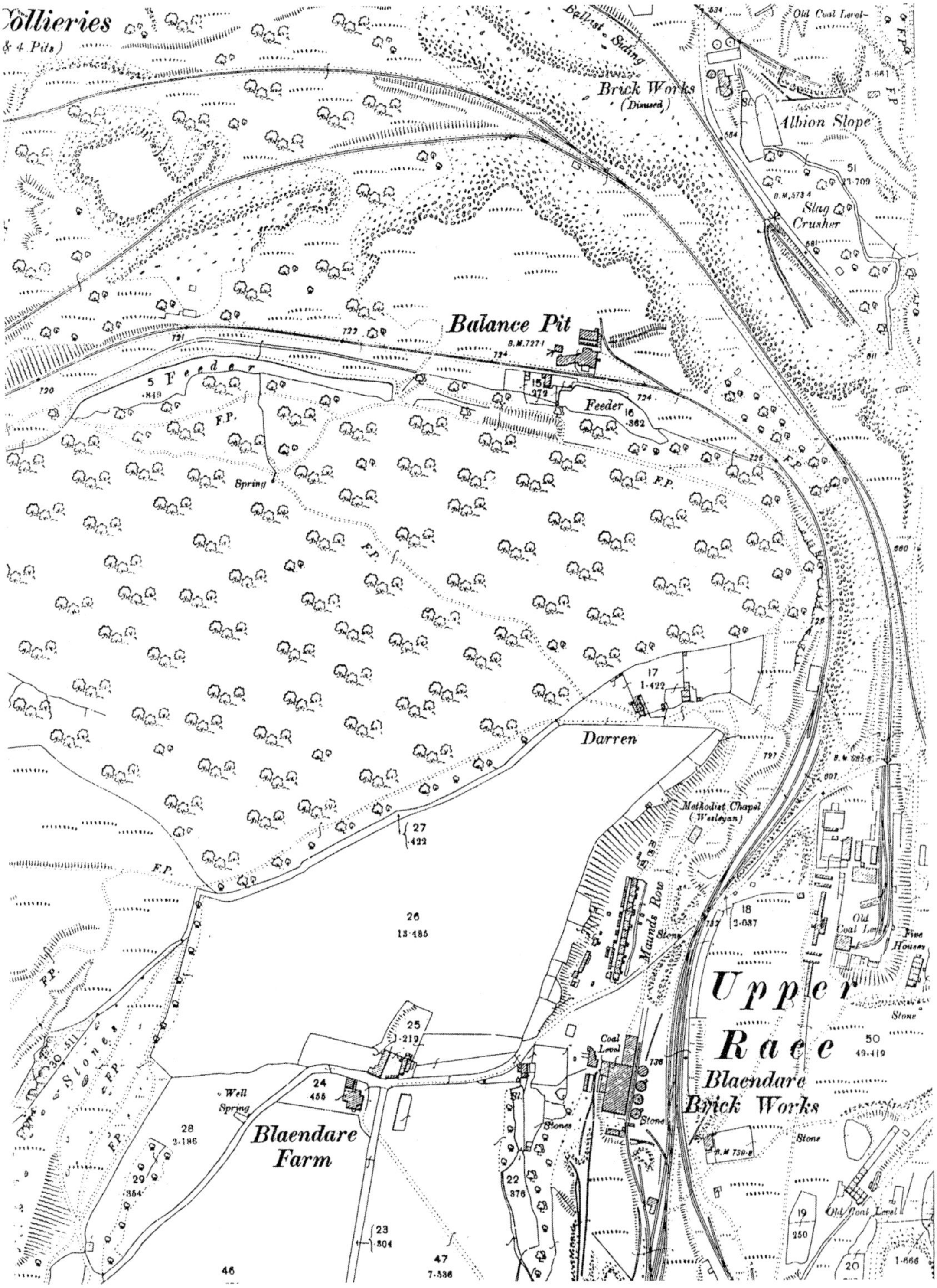


Fig 2: Map showing the Blaendare area, including the Balance Pits, Albion Slope at Lower Race, Blaendare Farm, workings at Upper Race, etc.

# Old Glyn Pits
## PRESERVATION WORK

Since 2002, work has progressed on the preservation of the Glyn Pits. As yet, only new galvanized roofs and some repair work to the walls of the vertical winder house have been carried out; while the pump house has been scaffolded in order to shore the walls, as a large quantity of dressed stone has been stolen from the rear wall.

We have now formed the Glyn Pits Preservation And Community Heritage Group, to coordinate work at the Glyn Pits. Our Chairman is Councillor Mr Neil Waite, while my colleague, Mr Clive Davies, is both Treasurer and Company Secretary. I am also a member of the group.

It was fortunate that demolition work at the Glyn was stopped before too much damage was done, because this mine is nationally important, and is the only one with the engines still *in situ.*

It is fairly certain that the vertical winder is the only one of its kind left, while the rotative beam engine, while not unique, is still rare, and again, possibly the only one still *in situ* in Britain, the mine being one of the first to use steam to both wind coal and pump water in Wales; and although they were at the latter end of it, they were, nevertheless, still part of the Industrial Revolution in Pontypool.

Fig 3: The large stone in the picture is 7ft 3in long, by 34in wide, and 14½in thick. It is some 34ft above ground level. All the masonry at the top of the building has now been re-bedded in mortar by a firm specialising in this type of work. On the corner of the building, painted black, is one of the four new angle-iron corner pieces which are linked together with tie-rods.

Fig 4: This shows the remoteness of the Glyn Pits site.

# SECTION 1

# 1
# Coal, its Origins and Chemistry

Recently, on a photographic visit to Big Pit Mining Museum, at Blaenavon in Gwent, one of the guides there, an ex-miner himself, told me that a young person, visiting the mine with a school party, had asked him: 'What was coal, and what was it used for?' This question, by a boy still at school, prompted the writing of this first chapter as an explanation of coal mining in general.

Coal fuelled the Industrial Revolution in Britain. The early iron-masters used charcoal, almost pure carbon, in their furnaces for smelting iron until Abraham Darby at the Coalbrookdale ironworks in Shropshire, developed the use of coke for this purpose in 1709.

The coal measures were laid down in the geological era known as the Carboniferous period by the swamp and land plants which first appeared in the geological period which preceded the Carboniferous, the Devonian. Well over half the world's coal measures were laid down in the Carboniferous period, and also during this period, the first fish appeared, the first amphibians evolved, and the first forests appeared, although they would not rival the forests we have today, either in size or type of fauna. Nevertheless, there were fine trees of a hundred feet or so. It is interesting to reflect that the peat bogs, some of which we are digging out for garden improvement today, are part of the coal measures of tomorrow! This period of luscious growth, the Devonian, gave the following period its specific character and name – the Carboniferous.

This period of rock formation began about 275 million years ago and lasted some 55 million years. A huge accumulation of vegetable matter, such as giant club mosses, tree ferns, and calamites (which were a species of giant rush), etc. occurred and became fossilised as the combustible black rock we know as coal. The greater part of the world's coal measures came from this age, with some also from the next period, the Permian.

In the islands adjacent to Spitzbergen and Bear Island, Devonian coal, as well as coal from the Jurassic and Eocene periods can be found: proof that this polar region has had a warm climate and supported large swampy forests.

These Upper Carboniferous deposits have played a great part in the economic development of mankind, especially in Great Britain and the United States, and this period is sometimes known as the Coal Measure Period.

These forests were able to thrive in this period of warmth and moisture for the best part of 20 million years, and the lands where they grew were subject to a general sinking, accompanied by repeated small changes of level.

The sinking was so prolonged that the total thickness of the sediments laid down along the shores piled up to well over ten thousand feet, while the changes of level drowned forest after forest where they stood, sealing their remains away under a sandy or muddy sediment, then raised the whole area above water again for another forest to grow, be flooded, killed, and preserved in its turn.

The results of these changes can be read today in the way the layers of rock generally lie in a coalfield, with the seams of coal alternating with bands of other rock like a Neapolitan ice cream. Often the layer below each seam is clay, formed of fine consolidated mud in which impressions of tree roots are frequent, and the lid over the seam is generally stratified shale or sandstone.

The actual process by which a mass of leaves, stems, and spores were turned into a hard black material like anthracite or house coal, 80 or 90% carbon, is a matter for the geologist or chemist. There exists every graduation between the hardest anthracite through house coal, cannel and brown coals to lignite and peat, which has scarcely more carbon in it than ordinary wood; meanwhile, the raw material for coals of every description is being accumulated today, mainly in the subtropical and tropical swamps and peat mosses.

What will happen to this material depends on the way it is accumulating, the sealing off it may get, and the weight and squeezing to which it will be subjected in the next few million years or so. It will never become anthracite, for example, unless it is subjected to the enormous pressure and temperature which exist deep in the earth's crust. There is plenty of other coal in the world besides that made by club mosses and calamites and other carboniferous plants.

According to an estimate made in 1913 by the Geological Survey of Canada, there are some seven billion tons of coal in existence. This contains four times as much carbon as the whole of our atmosphere today, and it is not known what effect the withdrawal of all this enormous bulk of carbon represented by these coal measures might have on the planet.

Coal is not a variety of carbon, but a mineral very rich in it. It is a compound or mix of compounds of carbon, hydrogen, and oxygen. Ordinary bituminous coal contains from 75 to 90% of carbon, 5 to 7% of hydrogen, and 5 to 20% of nitrogen and oxygen, plus small amounts of water, sulphur compounds, and mineral matter. In anthracite or steam coal, the oldest variety, there is 90 to 95% of carbon.

When heated in the absence of air, as in the retorts of a gasworks, and using condensing apparatus (Destructive Distillation) it can be shown that coal gives four main products. They are a mixture of gases – coal gas, ammonia liquor, coal tar, and a grey solid, coke. All these products also yield many by-products.

Bulk densities of house coal, anthracite, Phurnacite and Welsh dry steam coal average 50 cubic feet per ton. A ton of coal carbonised in a gasworks would yield about:

Fourteen hundredweight of coke
Three gallons of benzole
Ten gallons of tar
Twenty-one pounds of sulphate of ammonia
Seventy-two therms of gas

Purified coal gas is a mixture of 50% hydrogen, 35% methane, 8% carbon monoxide, and about 3% of nitrogen and oxygen. The remainder consists of compounds of carbon and hydrogen, such as ethylene, acetylene, benzene, and other hydrocarbons. It is considerably lighter than air. There are over two thousand by-products distilled from coal or its derivatives.

As coal forms, under the influence of extreme pressures and temperatures, the proportion of carbon relative to oxygen rises and water and other volatiles are driven out; these various stages are known as the ranks of the coal and in ascending order, the main ones are:

Lignite, or brown coal
Sub-bituminous coal, used in most power stations
Bituminous coal (the most abundant)
Semi-bituminous coal

Semi-anthracite
Anthracite, with a carbon content of between 92% and 98%

There are two types of coal recognised: humic or wood coal, derived from woody plant remains, and sapropelic, from finely divided plant remains, such as spores, and algae.

Average heating values for different fuels are given in British Thermal Units and one BTU is defined as the amount of heat required to raise the temperature of one pound of water by one degree Fahrenheit.

One BTU is equivalent to 252 calories.
100,000 BTUs = one therm.
Hydrogen = 60,000 BTUs per pound.
Oil fuel or petrol = (Average) 20,500 BTUs per pound.
Best Welsh coal = 16,000 BTUs per pound.
Average Newcastle coal = 14,900 BTUs per pound.
Lignite = 11,500 BTUs per pound.
Dry Peat = 10,000 BTUs per pound.
Coke = 13.000 BTUs per pound.

Finally, and depending on water content, the average BTU content of wood as a fuel is about one fifth to one quarter that of coal.

# 2

# An appraisal of the Old Glyn Pits and the surrounding area, near Pontypool, in Gwent

The Old Glyn Pits were originally known as Cwm-y-Glyn Colliery, and also as the Race Colliery. New Glyn Pits, or Nos. 3 & 4 pits, were at the bottom of the tram-road, on the site later occupied by Baldwin's Albion Road Colliery, or as we knew it, the Clog and Legging.

The Albion Road Colliery was driven in to the bottom of Glyn Pits up-cast shaft. The watercourse from the Old Glyn followed an almost parallel line, but from eighty yards down the up-cast shaft at the Glyn. This watercourse was above the New Vein and turned east near the Clog and Legging, to exit near the sawmills on Albion Road, close to the exit of the tunnel that extends from near the Drill Hall on Osborne Road in Pontnewynydd.

The Race Colliery, now the Old Glyn Pit, was sunk in the early 1840s on the orders of Capel Hanbury Leigh, whose initials, together with the date, 1845, are borne on a stone plaque over the door of the beam engine house. This mine was leased to a partnership in 1852, and to the Ebbw Vale Steel, Iron and Coal Company, then controlled by the Darbys, in 1855.

The Hanburys started their activities in Pontypool in the sixteenth century. There were furnaces at Trosnant, Pontymoile, Cwmffrwdoer, Town Forge, Osborne Forge, and also at Tintern, well known for its wire works etc., in the late 1500s. The name Tintern is significant here, because Terne plate was iron plate coated with Terne metal, a lead alloy with up to 18% tin and 1.5% antimony. It was used to coat steel to improve its corrosion resistance and working qualities. Terne plate was usually coated by hot-dipping with Terne metal, which acted as a carrier for, and as a lubricant in subsequent operations, such as wire drawing, as rust protection, as a paint base or to facilitate soldering.

The Old Glyn was preceded by the nearby Balance Pits, sinking of which commenced, again at the instigation of the Hanburys, in 1832. It was worked by water balance.

Five men were killed at the Old Glyn Pit on 23 January 1890. It was not known how the fire started, but candles, which were used at the Glyn alongside safety lamps, were thought to be the cause. This agrees with what I was told many years ago by an old friend, Mr Billy Watkins, whose father worked at the Glyn at the time of the tragedy, and who attributed the fire to the same cause. In order to extinguish this underground fire, a small reservoir at the rear of the pit yard was breached, allowing the contents, some 750 cubic yards, to flow into the up-cast shaft to flood the mine workings. Water was still being poured into the mine one month later, when arrangements were made to pump it dry. There were five men still in the mine, along with nine horses, and because of the poor condition of the mine after fire and flood, their bodies were never recovered, and so they rest there to this day. But one may be sure of one thing; that their comrades would have made heroic efforts to rescue them.

There is a commemorative stained glass window at St Cadoc's church, Trevethin, with the names included:

James Spear
Frederick Turner
Arthur Mills
Henry Price
David Lloyd

But what unfolded there on those fateful days was a double tragedy: to support the families as there was no social security at all in those days many of these men found temporary employment at the Llanerch Colliery. There, sadly, just days later, they were caught in the terrible underground explosion there, in a district known as Cooks Slope, so becoming part of the appalling toll of 176 men and boys killed there.

There were also three men killed here at the Old Glyn in 1864 by a small explosion. This had been caused, so it was said at the inquest, by an air-door being left open, thus affecting the ventilation system.

The Coroner recommended, in conclusion, that the jury should inform the mine manager that, in future, the air-door should not be left unattended and that a boy should be there at all times. So we still have a little lad of tender years in charge of something as important as mine ventilation. It makes one wonder just how many explosions and fires were caused over the years by such parsimonious attitudes.

I have personally known this venerable colliery, the only local reminder of a once great industry, since the year 1940, or 65 years, and have watched its steady but seemingly inexorable deterioration into its present state. It seems to me that, with a little foresight on the part of 'powers-that-be', the Old Glyn could have (and I firmly believe one day will) become a valuable part of the now booming tourist industry. It predates Blaenavon's Big Pit by some fifteen years. Equally importantly, the vertical winding engine housed here is the only one of its kind left in the world, apart from a similar one in the open air museum at Beamish.

Fixed on the north wall of the engine house is an enamel plaque, itself peppered with shotgun pellets, which states, in part, that: *This site is a monument of national importance.* So it is, but when one first looks at the site, and the dreadful condition these unique engines and buildings have been allowed to get into, the meaning of this statement may not be immediately obvious! This appraisal is dedicated to those who gave their lives to this dangerous industry and also to the exceptionally brave men of the Mines Rescue Service.

No 9.
THE GLYN PIT
PONTYPOOL

*From Mr. Jones, Manager.*

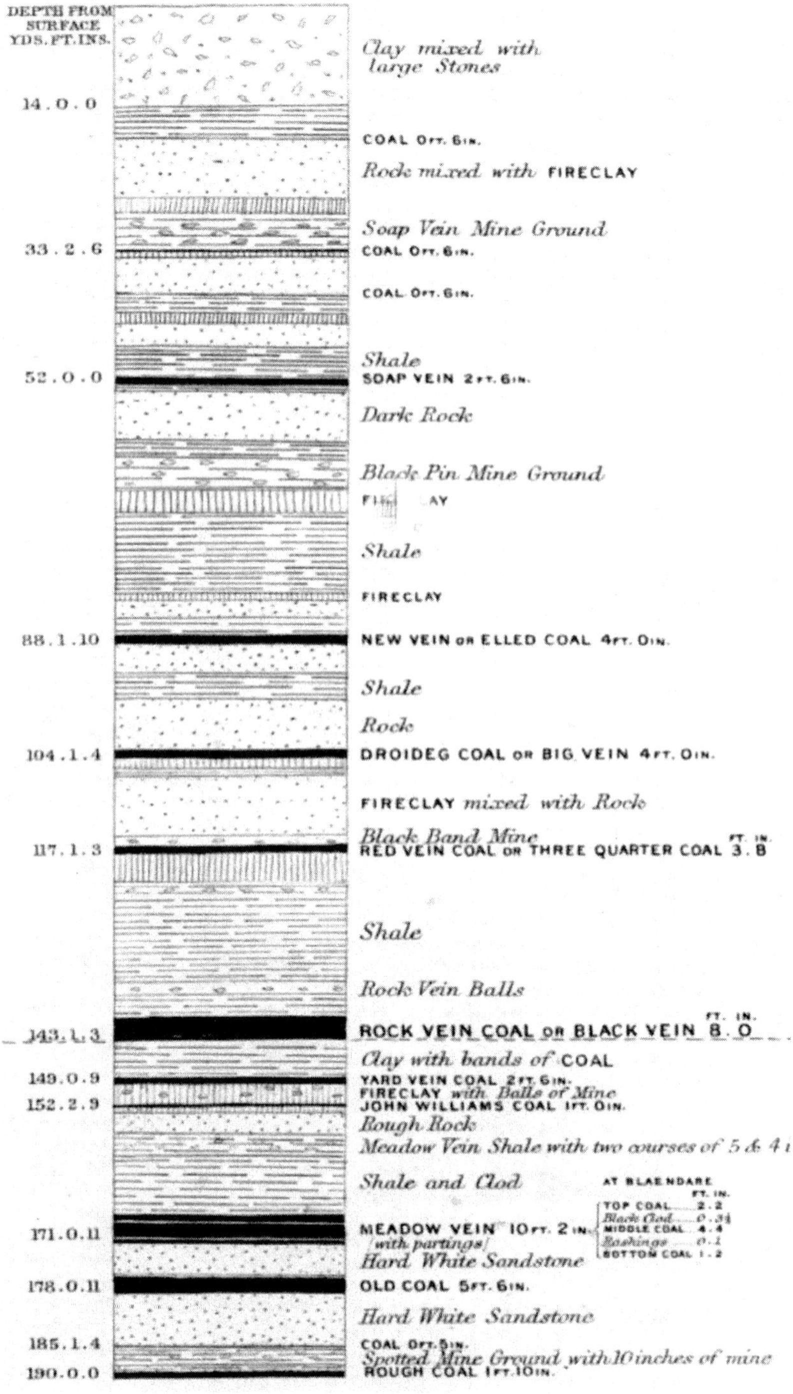

Fig 5: (Opposite). This is a section of the geological strata at Glyn Pits. This is taken from an engraving dated 1895. Along with various other strata, there is a total of around 42ft of coal in the several seams, the thickest of which, the Meadow Vein, is just over 10ft, followed by the Rock or Black Vein at 8ft, then the Old Coal at 5ft 6in. The other seams, such as the Droideg or Big Vein, the Red Vein, and the New or Elled Vein at around 4ft.

As will be seen from the section, there was also a large amount of fireclay and ironstone. The latter was also worked at the Glyn.

The Tillery Vein crops out at about forty-five yards above the Glyn Pits. This would be the seam worked at the colliery some three hundred yards or so west of the Glyn, Wern-Tillery, or Quarry Level (Wern or Gwern = Marsh).

Some of the terms used in this section:

| | | |
|---|---|---|
| Balls of Mine | = | Nodules of Ironstone |
| Bastard or Bast | = | Impure |
| Bind | = | Shale |
| Blackband | = | Ironstone |
| Clift or Cliff | = | Shale |
| Clod | = | Clay or Fireclay |
| Mine | = | Ironstone |
| Pennant | = | Blue Feldspathic Sandstone |
| Pin | = | A band of ironstone |
| Rashings or Rashes | = | Shales with streaks of Clay |

Fig 6: The Old Glyn Pits at Pontypool; it is not dated. The colliery tram in the foreground is on the bottom of the incline to the screens, which were about 100 yards away to the right. The headgear seen here was said by a Coal Board engineer to be '50ft to the centre of the wheel bearings.' We measured the height to the eaves of the winder house as 34ft, which makes his estimate seem a little low as it is near twice that height. The original boiler house was near where the transformer house is here, with the boiler stack behind it. Behind the headgear, the blacksmith's shop and tram repairer's shop can just be seen, while in the left foreground are some steel roof support rings. The coal wagons on the left are supplying coal for the Lancashire type boiler which was installed just off the left of the picture. One can also see the top of the air seal around the up-cast shaft over the transformer house.

# 3
# THE VERTICAL WINDING ENGINE

Unlike the rotative beam engine at Glyn Pits, we have no documentation or date yet for the vertical winder. We know from an inventory taken in 1852 that the beam engine was then the only engine on site, and from a drawing in the book *Le Tour du Monde*, published 1865, that the vertical winder was on site then, so we have a date somewhere between those two.

It is a house-built engine, in that it is part of the engine house containing it. It had manually operated Cornish double-beat equilibrium valve gear, and from the diameter of the winding sheaves, we know it made around thirteen and a half turns from bank (pithead) to pit bottom.

The cylinder bore, measured from a used cylinder gasket found in the engine pit, is about – one cannot be precise here as to the bore – 36 or 37in, with a stroke of 5ft 6in.

It used flat wire winding rope, much of which is still on site, on reel type drums; that is to say, the rope wound on itself, rather than across the drum, as in later types of engine. This arrangement gave no fleet angle (fleet angle being the angle between the rope and headgear pulleys caused by the sideways movement of the rope across the winding drum) so the headgear pulleys at the Glyn carried a flat groove slightly larger than the rope width – as opposed to the normal deep vee type groove for a round rope. It also allowed for a narrower rope exit slot in the front of the engine house, and, here at the Glyn, the slot in the roof for the down-cast rope.

When a cage was at bank, i.e. at the pithead, the diameter of the rope on the corresponding sheave was larger than that of the unwound rope sheave on the opposite shaft, i.e. the full cage at pit bottom, and because of this greater diameter, it gave a certain mechanical advantage to the empty cage at bank over the full cage at pit bottom and the weight of its rope hanging down the shaft. This helped to balance the winding cycle, making it easier for the winder to start.

In later winding gear, imbalance was catered for by installing the cylindro–conical scroll type of drum, in which the drums, one for each rope, were cone shaped, with the scroll track for the rope winding up each drum, so that as the rope wound down the scroll on one drum, it climbed the other, giving a continually varying diameter as the wind progressed.

Fig 7: (Opposite). Shows hand-operated valve gear, with Cornish double-beat valves. The vertical lever in the foreground is the brake lever. In the other part of the building, under the winding sheave, was a small temporary steam haulage engine serving the down-cast pit. I believe this haulage engine is now preserved in Caerleon, Gwent.

Fig 8: The piston rod, lower right, attached to the cross-head and connecting rod, top right. The hand-operated valve gear is shown clearly, with the hand lever, left, and the two valve spindles under the operating levers. The fluted columns of the engine frame can also be seen. In this picture, all the bearing caps have been stolen from the valve gear. Today, the valve gear is missing completely as is one of the small-end bearings on the cross-head. Photograph courtesy of John Cornwell.

Fig 9: The cross-head guide bearings. These are tapered in order that they may be adjusted with the four adjusting bolts just visible at the top of the brasses. The two set-screws in the centre acted as stops, to prevent the bearing being inadvertently tightened too much. The brass on the left has an extra piece inserted against the guide bar. This does not appear to be to make up for wear, but to make up the difference between the size of the brasses and the guide bar at installation. The guide bars themselves could be adjusted by inserting washers behind them.

Figs 10 and 11: Two views of the main and secondary cranks. Fig 10 shows, left to right: secondary crank driving depth indicator gear; centre, drag-link, which allowed for axial misalignment of the shafts; centre, top of connecting rod. This connecting rod is 18ft in length and 8in thick at the widest part. On the right can be seen the main engine crank, with the main engine bearing and rope-wheel to the extreme right. The great size can be judged from the scale ruler, which is 2ft. Fig 11 shows the detail of the drag-link end.

Fig 12: Shows the rim of the brake wheel. The sections of this wheel, which are on the same side as the main crank, have a hollow rim for balancing purposes. This is the opposite side of the wheel with 'bolt-on' extra weight, which was made for the purpose. With a wheel of this size and weight, balance was very important, even though the RPM was low.

Fig 13: Shows cutaway section of supporting timber. This was done to accommodate the throw of the connecting rod.

Fig 14: The guides for the flat winding rope, with the down-cast rope in place. The up-cast rope, in far better condition, can be seen to the right.

Fig 15: The bearing plummer blocks for the beam on the beam engine. The plate dropping at an angle from the bolt in the foreground was part of a support for an oiling platform – now rotted away – for the beam bearings and connecting rod bearings. The scale ruler is 2ft.

Fig 16: The wood and wrought iron rope-guide horns on winding sheaves. This is the down-cast rope-wheel.

Fig 17: The up-cast rope. The difference in condition is clear. The up-cast rope was originally carried over the head-frame and connected to the cage on the up-cast shaft, and it appears it was dragged into its present position by the unbalanced forces of the down-cast cage and rope in the down-cast shaft, which caused the engine to turn over. This probably happened when the up-cast cage was disconnected from the flat rope for use on the round rope of a smaller temporary winding engine.

Fig 18: The method of shortening flat winding rope by inserting pieces of flat timber between the coils as the rope was wound on the reel. This method saved disconnecting the rope from the cage for recapping, which was a hard and time-consuming business. The method of shortening shown here was only applicable to flat ropes, which in any case were far more prone to stretching than round rope. It can be seen that the piece of timber has been nicked in two places to enable it to conform to the shape of the rope-wheel.

Fig 19: The inner side of the wheel nearest the up-cast rope-wheel, showing the method of construction. The brake band is 5 1/4in by 1/2in. The bolts were for securing the brake blocks, which were burned off when vandals set fire to the top floor and roof. The small loop in the centre of the brake band was a support for the band. It was connected to a roof timber. The piece of rusted iron was a rope-guide.

Fig 20: The rim of the wheel. It measures 11in by 8in, and the bolts securing the sections are 1 1/4 in diameter. The scale ruler is 2ft.

Fig 21: The side and below view of the engine cylinder. This cylinder is 42in overall diameter and 36in bore. (The bore taken from a used cylinder cover gasket.) The engine guide bars can be seen mounted above it. This cylinder bears similar stiffening rings cast around it – just visible below the gussets, top – to the beam engine cylinder.

Fig 22: One of the cast iron seatings for the headgear backstays.The bottom half of these was cast as part of the iron plate on which the scale ruler rests. The immense size of these backstay timbers can be judged from the length of the bolt fitted through the near casting; the winding rope for the up-cast shaft exits through a slot on the far side of this casting; the winding rope for the down-cast shaft exited through a slot in the hip roof. The other two seatings can be seen. The quality of the masonry speaks for itself.

Fig 23: The overhead winding sheaves at Glyn Pit, with the author standing next to them. This emphasizes the great size and heavy construction. The brake band – complete with some of the wooden brake blocks – can also be seen. Unlike more modern post-type brakes, this type uses a band around the wheel; it was in two halves, and is identical to those used on most water balance wheels. It is explained more fully in the section on the vertical winder. The crank nearest the camera is the secondary crank, and drove, in this case, the depth-indicating gear. It is connected to the main crank by a drag link, which can just be seen between the two cranks; this eliminated the need to line the two shafts axially.

Fig 24: The classic one of the Glyn Pit, the rotative beam pumping engine.

Fig 25: As will be seen, this is a house-built engine, i.e. the main structural members are built into the engine house. In the far top corner can be seen some of the brake levers. Also clearly seen is the jawed-type connecting rod, fluted columns of the engine frame, and what remains of the hand operated valve gear. This engine is now badly affected by rust, particularly the piston rod. The cast iron girder built into the wall over the engine is 12ft long, 14in deep, and 3ft wide. It will also be noticed that one of the big-end bearings on the cross-head is missing.

Fig 26: The lower set of valve gear on the vertical winder. The 12in engine exhaust pipe can be seen extreme left, with a supporting bar hanging from the underside of the engine-room floor. This exhaust apparently leads to the brick stack at the rear of the engine house. The main engine cylinder is on the extreme right; note the stiffening bands around the cylinder. By following the fluted support column up, one can see the link bar from the upper set of valves. There is a similar support column at the rear of the valve adjustment rods.

Fig 27: The vertical winding engine braking mechanism rim of the brake wheel. As can be seen, these sections are cast hollow; they are placed opposite the main crank and connecting rod, and serve to help balance the engine. The flat winding rope, left, was for the up-cast shaft, and exited the engine house through the slot in the masonry on the west wall. Also to be seen are the palms cast on the spokes of the brake wheel to carry the inner reel type winding wheel which is partly of wooden construction. The casting on the rotten beam behind the wheel was the anchor point for the brake bands.

Fig 28: The main shaft of the vertical winding engine protruding through the south wall of the engine house. Both the bearing cap and the bearing brass are now missing. The size of this shaft etc. can be judged by the 2ft scale ruler. The bearing is fixed to a cast iron plate which is built into the wall. This, in turn, is secured by 12 long holding down bolts, which are some 2in thick and 30ft in length. They are of square section and are built into the interior of the masonry. They are secured at the bottom by plate and cotter fixings.

Fig 29: Shows one of the many wooden brake blocks attached to the brake band.The blocks are 18 x 6 x 4in. The immense size of the cast iron wheel rim can be judged by comparison with the 12in scale ruler wedged under the brake block.

Fig 30: Main engine bearing of the vertical winding engine. The main crank is on the left, with a throw of 33in, giving an engine stroke of 5ft 6in. The hub of the brake wheel is on the right, secured by iron wedges; these in turn are secured by the band bolted around the hexagonal main shaft, which is 14in across the flats.

Fig 31: Close-up of dressed stone (ashlar) on the vertical winding engine house. This is typical of the whole of the building, and is continued down to basement level, and even inside the six pockets for the holding down bolts. This small area is just inside the broken brickwork used as an entrance now.

Fig 32: The stack has long since been demolished, and the roof of the engine house mysteriously caught fire some years ago. Level with the arch in the wall are six square holes. These correspond to six pockets in the basement, where the anchor points for the holding down bolts are. Just to the left of the main house can be seen other buildings. These were a small pumping station and pressed steel storage tank. Water from the Glyn itself, Glyntillery Colliery, and Quarry Level Colliery was pumped into this tank, before being pumped, by the electric pumps in the smaller building seen here, to the reservoir at the Upper Race, also now disused. Most of the steel water mains are still in place under the route of the railway track to Blaendare, and to the Glyn from Quarry Level.

Fig 33: Shows the cast iron plate built into the wall of the engine house. This plate supported the three cast iron brackets, which in turn supported the backstays for the headgear. These were supported in turn by the three stone corbels, at the bottom. The slot where the plate has been cut accommodated the up-cast winding rope and was original. The plate itself has been crudely cut by drilling, and from the angle of the drilled holes, it was done after installation of the plate etc., perhaps to render rope-changing easier.

Fig 34: One of the palms cast on the brake wheel to carry the inner winding sheave. This sheave was for the up-cast shaft. The part wood (elm) construction of this sheave can clearly be seen here, as can part of the woven winding rope. Note the massive size of the castings.

Fig 35: Brake wheel, inner rope-wheel, partly of wood construction, bolted to the brake wheel, with outer, cast iron rope-wheel behind. A theory has been advanced that the inner wheel is of later construction; in effect, an add-on accessory. As will be seen from the picture, the brake wheel was designed for this wheel to be bolted to it from inception (see text). The brake band can also be seen clearly, although many of the brake blocks were burned off in a fire some years ago. The brake adjusting bolt for the upper band is on the right.

Fig 36: Auxiliary shaft driven by secondary crank, right, which is driven by a drag-link from the main crank. This method of driving it minimises the need for too-precise alignment, especially, as here, when the bearings were on a wooden base. The pair of crown wheels, left, drove the depth-indicating mechanism. It has now, due to rotten timber support, fallen out of mesh. The purpose of the chain sprocket is not clear, but it could easily have been used to drive a valve mechanism. Indeed, this has been done on some of the other vertical engines. It is probable that this whole mechanism was part of an engine package, as it seems rather too heavy and substantial for such a light duty. The other end of this shaft is depicted on Fig 38, together with its bearing, also missing its top brass.

Fig 37: Anchor point for flat winding rope. This one, on the outer wheel, is for the down-cast shaft.

Fig 38: Chain wheel and chain of depth-indicating mechanism. This chain carried stops which contacted single-stroke bells at the appropriate time, to indicate to the engine man the position of the cage in the shaft. These two bells, of which only part of one remains, were fitted on the engine-house wall in front of the engine man. As with everything at the Glyn Pit, note the substantial nature of the equipment.

Fig 39: Adjustment thread and nut for the lower band.

Fig 40: Adjustment thread and nut for the upper brake band.

Fig 41: One of the rope-guides on the winding sheaves with some of the rope still in position. This was for the down-cast shaft, and there appear to be no iron linings to this guide. Note also the condition of the nearer, up-cast rope in comparison. It appears relatively new.

Fig 42: View of cross-head slipper bearing, showing adjusting set screws and small blacksmith-made pointer. This would have been fitted to enable the engine man to judge the position of the crank, in order not to stop the engine on-centre from where it would be difficult to restart. The slipper brasses are tapered to allow adjustment.

Fig 43: Shows the condition of some of the wooden beams at the Glyn. The top beam in the picture supported the main bearing plummer block for the secondary shaft. The condition is obvious. The other beams are in similar condition. Note the cast iron bracket – under the 2ft scale ruler – tying these two beams together. The twin beam to this one, out of camera shot to the right, supports the other end of the secondary shaft, which terminates in a crown wheel. This meshes with a smaller crown wheel driving the depth-indicating gear. These wheels have now dropped out of mesh due to the rot in the supporting timbers (see the photograph below).

Fig 44: This shows the method of incorporating holding down bolts into the masonry. This is the cast iron plate between the two left-hand corbels, with the rope slot lower right. Note the crude way in which the plate was cut by drilling. Under the circumstances, and given the thickness and width, this was probably the only way in which it could have been cut. These corbels supported the headgear backstays. (Scale ruler 2ft long.)

Fig 45: Rim of brake wheel. This wheel is 18ft diameter. The hollow sections in the picture are on the same side as the main crank and connecting rod and are designed to help balance the engine.

Fig 46: The flat winding rope on the down-cast rope-wheel and rim of the brake wheel. Note the wooden strip wound in with the rope. Because this type of rope was so prone to stretching, this procedure was sometimes carried out to shorten the rope in order to bring the cage back to the correct level with the pithead or bank. It was not then necessary to unwind the rope for shortening at the anchor point of the wheel, which was a time-consuming and difficult business. This would have been done when the cage was at pit bottom, and the rope unwound. Looking closely, one can see the wire used to attach the wood strip to the wood of the wheel while the rope was wound over it, also – bottom left – where it was twisted to tighten it. The hollow rim of this part of the brake wheel can be seen clearly here too.

Fig 47: These levers applied the brakes to the brake wheel. Note the 2ft scale ruler. The rod dropping from the near end of the bottom lever was for the driver's hand brake. The two flat bars, dropping from the centre, were for a steam slave cylinder. These were arranged to apply both brake bands (top and bottom) simultaneously.

Fig 48: Connection of braking levers to brake bands. It can be seen that the top lever pulls the top band down, while the bottom lever pulls the bottom band up.

Fig 49: Top of the connecting rod, and vertical winding engine main bearing, left secondary crank, and auxiliary shaft on right. This shaft was used in this case to drive the depth-indicating mechanism, seen on the far wall, but could easily have been used to drive the valve gear under other circumstances. Note the one crown wheel – right – has dropped out of engagement due to rot in the supporting timbers.

Fig 50: Shows guide bars and slippers. The slipper bearings are tapered to facilitate adjustment. Of the six adjusting bolts, top, the four outer ones are for tightening the bearing, while the two inner ones – set-screws – act as stops, to prevent over-tightening. Note the extra piece of brass inserted on the left-hand slipper. This appears to have been fitted not as a compensating piece for wear, but to bring the bearing size to that of the guide bars. This would save the work of moving the guide bars inward, although they were adjustable. Looking carefully, one can just see a hexagon nut to the left of the top of the scale ruler holding the guide bar to the frame. These could have been packed out for adjustment with plates or washers.

Fig 51: A larger view of the winding rope at Glyn Pits. As before, notice the exteremely crude method of terminating the rope.

Fig 52: Shows a spare section of rope which was left behind the winding sheaves and across the main shaft of the winder. It has been cut in the past and the individual wires peeled back, probably by someone to illustrate the method of construction. This is now clearly shown. This section appears to have been unused, and was probably an off-cut from rope-changing.

Fig 53: Close-up of end of the winding rope.The woven flat rope, which was $4^1/_2$in, has stretched down to just 3in. This was the end of the up-cast rope, drawn back over the headgear to the rope sheaves by the weight of the rope in the other shaft overbalancing the engine when the up-cast cage was disconnected for use on a temporary smaller haulage engine.

Fig 54: The end of the flat winding rope. As will be seen, the frog at the end of the rope has been worn right through. In this frog was an iron ring, to which were connected the four chain bridles supporting the cage. The crude method of joining the frog to the rope will be evident. The frog itself is of wrought iron, blacksmith-made, probably at the pit smith's shop. It is as well to remember that men were also wound on this rope.

Fig 55: Top of up-cast shaft, showing two (of four) keps. These were pushed aside by the ascending cage, then fell back into position to support the cage in the correct position – in line with the tram rails – while at bank. When the cage was to descend, the banksman pulled them clear with a hand lever. Depending on where one was in the coalfields, they were also known as 'legs', 'shuts' or 'fans'. There are two more on the opposite side of the pit top, but due to the levers being cut, these have fallen to a horizontal position.

Fig 56: Boiler or air receiver buried near the vertical winding engine house. I do not know if there is only the one or more, but they are set on brick seatings. I also think it unlikely they are boilers, because there is a 10in pipe taken from near the shell of this one about three-quarters of the way up, and without a valve, so that steam could not be drawn from it without the risk of severe priming. Whether the brick stack at the rear of the engine house is connected to these is hard to tell without a more thorough examination, but there are neither flues nor boiler uptake in evidence.

# 4
# Boilers etc. at the Glyn

Steam was generated at the Glyn by two Cornish boilers, working at 50 psi. Each boiler was 26ft 6ins long, 7ft diameter and, being Cornish, had a single fire-tube. This was usually about six-tenths the diameter of the shell of the boiler. The two boilers at the Glyn were flued to a separate stack, which is shown on older photographs.

Some Cornish boilers were also fitted with Galloway tubes. These were smaller cross-tubes, usually three or four in number and they were made conical both to improve circulation and also so that they could be fitted into place. They were fitted in the tubes at an angle of about 45 degrees. Their purpose was to increase the heating surface of the boiler, and to improve the water circulation by lifting water from the bottom of the shell by convection.

As well as the usual boiler fittings of pressure gauge, water level glasses, safety valves, blow-down valve, etc. there would also have been a fusible plug. This plug, of bronze or gunmetal, would have been fixed in the roof of the firebox, right over the fire. It was bored through and the hole was then fitted with a bronze plug secured with either white metal or a tin–lead mixture, which, if through low water level it became uncovered, would melt, allowing the centre plug to be blown out, admitting steam and water to extinguish the fire.

There was also, at a later date, a Lancashire boiler installed; this too is to be seen on some older photographs. It was on the north side of the engine house and it appeared, judging by the building housing it, to be temporary, and also connected to the stack at the rear of the winder house. The original boiler house was demolished to make way for the brick/concrete transformer house which is still in place.

This supplied power to a small electric haulage engine on the temporary headgear over the up-cast shaft, and also to the pumps in the shaft, which were not at pit bottom, because, apparently in later years, both shafts flooded up to the level of the original watercourse, some 80 yards down, and it was here, in what was called a 'lobby', that the old Pontypool Gas and Water Company obtained water for the Upper Race reservoir. Water was pumped up to a large steel tank near the pithead, and from there to the Upper Race.

In recent years, there were two main types of boiler for steam raising; one was the fire-tube boiler, the other the water-tube type. As an example of the fire-tube type one only has to look at the railway locomotive boilers, fairground or showman's engines, or the Cornish and Lancashire boilers. In this type the tubes are surrounded by water, with the fire and hot gases passing through the inside of the tube. Water tube boilers – such as the Babcock & Wilcox marine boilers – have the fire outside the tube and the water inside. Both types can be fitted with superheaters.

Superheaters, as the name implies, superheat the steam leaving the boiler. Steam in contact with its water is said to be saturated steam and saturated steam at a given pressure can exist at only one temperature and abstraction of heat slowly causes some of the steam to condense.

Put simply, saturated steam will condense relatively quickly, whereas superheated steam can give more work before condensation. In fact, the more one puts in, the more one gets out. In general, water-tube boilers are more suited to the generation of steam at very high pressures. The efficiency is about the same for both types.

The Cornish type of boiler was invented by Trevithick in 1812. Before this, boilers were of the beehive or haystack types; neither of which was suitable for the high pressures – in this case about 50 psi – needed for efficiency, because the higher the initial pressure of the steam, the more work can be got out of it.

This type of boiler was vastly improved on in about 1840 by the invention, by Fairbairn, of the Lancashire type. This had two fire-tubes, each about one quarter of the size of the boiler shell, which was anything from 25–30ft, and up to 9ft in diameter. The furnace was placed in the front of the fire-tube, and was usually around 6ft long.

One of these boilers, a Lancashire type, was installed at the Glyn in later years; it was placed to the north of the winding engine house, and connected to the brick stack which was already there. Boiler fittings for later type Lancashire boilers were:

A junction valve for steam supply;
A dead-weight type safety valve;
A low water/high steam valve.

This latter valve was designed to blow off if the water fell below a safe level, or if the steam pressure rose too high. On the boiler front were a water feed valve, pressure gauge, and two water gauge glasses, which indicated the level of water in the boiler. These were fitted with automatic cut-off valves, which, if the glass broke, cut off the steam supply to the glass. This was accomplished by bronze balls, which would be carried by escaping steam onto seatings which would then cut the steam supply off. Finally, below the front floor-plate was the blow-down valve. Its purpose was, among other things, to flush out of the bottom of the boiler any sludge collected there. This was controlled by a long T-shaped handle, which, when fitted to the valve, projected up in front of the furnaces, so that it was in the way of firing the boiler. Also for safety reasons, this handle was fitted with lugs which engaged with slots on the valve body, so that it could not be removed with the valve in the open position.

Boiler feed water was usually taken from the hot-well, which was supplied with hot water by the condenser and sometimes by a feed water heater.

In one of the local tin works, the Town Forge, in Pontypool, there was a 'Green's economiser'. This device used boiler flue gases, which would otherwise be wasted, to heat feed water for the boilers. It was distinctive in having a small engine on the top, in those days usually steam. This had an automatic reversing device on it and its purpose was to move scrapers up and down the outside of the water tubes to keep them clear of soot etc.

At the Glyn Pit, there is a small reservoir at the rear of the pit yard, which supplied the condenser cold-well as well as the boilers with water. An examination of the cold-well today shows that the supply from the reservoir has been cut and the hole in the tank patched over, and the exhaust pipe from the engine has been taken out through the side of the building to a short vertical exhaust pipe there. A T-junction has been used, and it has not been disconnected from the condenser. I have not been able to find out when this was done, as even a few years at a mine would make a great difference.

The drain from the condenser cold-well/pumphouse is at the side of the tram-road on the approach to the Glyn. It is a small stone-built tunnel on one's right, now obscured by undergrowth.

# 5
# Winding ropes

The first rope in which metal wires were used was invented by Andrew Smith and a J. L. Hood, in 1836, and was known as the 'Selvedge Rope'. In this rope the wires of the strands were laid together and bound with yarn or twine. These strands were then laid side by side, just as in flat ropes, and they were parallel with the pull on the rope. The great problem with this method of construction was, that as soon as the twine or yarn wore, the rope became little more than a bundle of loose wires. In 1840, a Mr Newall patented a machine by means of which the wires were laid into strands without being twisted on their axis. This gave us the term 'lay', which is commonly used instead of twist. Then, in 1857, one Rolbling constructed a rope which was built of three strands laid in one direction, and three in the other, the idea being to prevent the rope twisting on its axis when loaded. In 1879, John Lang invented a rope which he claimed to lay all the strands forming the rope, and all the wires forming the strands, in one direction, with the object of forming a rope capable of greater resistance to wear and tear, which would outlast others constructed until then.Then in 1884, T. C. Batchelor invented a rope in which a section of the outside wire enabled them to be locked together, giving them the name 'locked coil' which gives a smooth outer surface with a large wearing area.

The winding rope at Glyn is of woven steel construction (see Fig 52/53) but which type of steel I do not know. It is 4½in wide and ⅞in thick, and according to enquiries I have made, weighed, depending on whether it was steel or iron, up to 15½lbs per yard. There were many types of woven winding rope in use at this time, ranging from this size down to the smaller size of 2¼in by ½in 'plough steel' with a safe working load of 80 cwt or four tons. Also there was 'best crucible steel' rope, 4½in by ¾in, with a safe load of 180 cwt, or nine tons. This type of rope is very prone to stretching, and on examination now it will be seen that the rope at the Glyn has been stretched well beyond present day safety standards, being barely 3in wide in parts, and less than ¾in thick, although not necessarily weaker (see Fig 52/53). But when one examines the termination for the up-cast cage closely, it can be seen that the sharp edges of the broken frog would soon have cut into the rope. Referring to the photograph, the method of terminating the rope at the Glyn can be seen to be extremely crude, with five pairs of iron clamping bars, approximately 12in x 4in x 1in, held with 1⅛in bolts. These are fitted on the last 12ft or so of the rope, while the frog itself is of wrought iron, folded and forge welded into a loop. In view of the foregoing, we may conveniently step back in time a little at this point, to Christmas Eve, 1873, in fact, when a report in the *Usk County Observer* tells us that:

> There was a serious accident at the Old Glyn Pit on Christmas Eve, when a winding rope broke, precipitating the cage and its tram of coal to the pit bottom. Before this happened, there were eight men waiting to ascend the pit in the bond, and several men waiting to descend. These were told to wait by the Banksman while a tram of coal was wound up first. This had just reached Bank [the pithead] when the winding rope broke, allowing the cage and tram to descend to the bottom of the shaft with a loud noise, and it will take some days to ascertain what damage has been caused.

This would have occurred in the down-cast shaft; that is, the one nearest the vertical winding engine; and bearing in mind the condition of the winding rope in the accompanying photographs, it is not surprising.

There was a smaller winder and steel headgear installed at a later date on the up-cast shaft for pump installation and servicing. This would have used round rope, and I think it safe to assume that they would have used the original up-cast cage by disconnecting it from the flat rope, allowing it to rest on the 'keps' (also called 'fans', 'legs'or 'shuts' depending on where one was in the coalfield – (see Fig 55) which would have caused the winding gear to become unbalanced, thus the rope in the down-cast – some 190 yards at 15lbs per yard, plus the cage on the end (which would have fallen into the sump) – would have been sufficiently heavy to start the vertical winder turning over, and dragging the rope up over the headgear and down to where it is today. I see no other reason for the up-cast rope to be in its present position. If this is the case, then the other cage is still at pit bottom.

A feature of the Glyn not seen in many collieries was that winding took place in each shaft alternately, as there was only one cage in each shaft. I don't know whether coal was wound in both shafts because using the up-cast shaft would have meant breaking the air seal every time a cage came to bank. But from one old photograph, there appear to be flap doors which would be lifted by the ascending cage. On this photograph, the top of the pit is not sealed completely around the headgear, as it was later. The shafts at the Glyn were 190 yards deep to the sump. Winding took place from 180 yards. The pithead gear was of pitch pine, and was, according to a Coal Board engineer, '50ft high to the wheel spindles'. If one compares this headgear on old photographs of the Glyn to the vertical winding engine house, which is 34ft high to the eaves, then this estimate is far too low. This same engineer has also shown both winding ropes as brought out through the front of the engine house, and this too is incorrect, as one rope, the down-cast, was brought out through a slot in the roof. This structure was replaced in the 1950s by a smaller steel headgear, with an electrically driven haulage engine. There was also a small steam haulage engine mounted under the original winding sheaves. This engine has now been preserved by a far-thinking person in Caerleon, Gwent.

# Section

# 2

# 6
# THE BEAM PUMPING ENGINE

The maker of the beam engine at the Old Glyn Pit was probably the Neath Abbey Ironworks, and this seems to be confirmed by the extant records, but as with the vertical winding engine, the beam engine itself carries no maker's name.

The stone plaque over the door to the pump house carries the inscription:

C.H.L.
1845.

These are the initials of Capel Hanbury Leigh, and I see no reason to doubt that assertion, since in the archives for the Neath Abbey Ironworks, there is a record that a 30in engine for Mr Hanbury was ordered.

The engine itself is a single cylinder, double acting, condensing rotating beam engine with a 6ft stroke and a bore of 30in. Watt's parallel motion was used on both piston and air pump rods. The valves, driven from a single eccentric on the first motion shaft, are of the Cornish double-beat, equilibrium type. The palm of the eccentric is now missing, and part of the valve operating rod, of wrought iron, has also been burned off; only the strap remains on the eccentric.

This valve rod, which travels under the floor, has an inverted U forged in it. This is arranged to drop over a roller fitted to the valve operating spindles, and could have been lifted out of engagement from the driver's position. This roller, about 2½in diameter (see Fig 76), is part of a hand-operating mechanism. In one document I read some years ago, this was described as a gab slot or lost motion. I fell into the classic trap at that time of taking someone else's work as correct, and not checking for myself. It was not correct. Therefore I could not see how, with just a single eccentric, this engine could be reversed for winding. It was on a later trip to the mine that I noticed that there were lost motion slots on the vertical valve rods. This would have made them unnecessary on the horizontal rod, and this was the answer. I then took a photograph (Fig 76) which made the whole thing clear to me. The valves were hand-operated for winding duties, and automatically driven by the eccentric for pumping.

Attached to the mechanism beneath the floor is a short vertical lever, the end of which has been narrowed slightly. This was to accommodate an extension lever, which brought it within comfortable reach of the engine man who could, when need arose, reverse the engine and also, using the valves to reverse the steam flow, stop the engine.

There is a certain amount of lost motion on the vertical valve-operating levers. This would have been necessary in order to lower the valves onto their seats, as the motion from the eccentric would have exceeded that necessary for valve lift. In these vertical operating levers are two clevis joints. These have had small wooden wedges driven into them in order to take

Fig 57: Beam pumping engine. Under the cast iron beam can be seen two of four holding down bolts for the beam bearings. These are similar to those used in the vertical winding engine house. The plaque between the bolts carries the initials C.H.L.1845. The vertical connecting rod is of oak, with the remains of the horizontal connecting rod sticking up from the ground in the foreground.

up some of the wear (see Fig 58/59). These levers are very badly corroded now, and will not last much longer. All we could do as a temporary measure was to pour several gallons of sump oil on them. This does seem to have helped.

Mounted on the first motion shaft is the flywheel. This is some 17ft in diameter, and is built up from cast iron sections. There are eight of these in the rim, 6in x 7in, and each has a number of barring slots cast in it. These are necessary on a single-cylinder engine for starting purposes because, as with the vertical winder, stopping on centre, either top or bottom, would have made the engine difficult to start; but by inserting a bar into one of these slots, the engine could be barred over until the piston was over centre, when starting could be accomplished easily.

The overhead beam of this engine is driven, by Watt's parallel motion, from the vertical piston rod. The other end of the beam drives, via a wooden connecting rod, the main engine crank on the first motion shaft. Also on this is mounted the single eccentric, the flywheel already described, and the smaller of two large gears. One of these is near 4ft diameter, with the larger gear, which is fitted on a shorter shaft, being near 9ft. Both gears are just over 1ft wide and have a pitch of 4in. Also on this shaft is the secondary crank which drove the horizontal connecting rod, also of wooden construction, which in turn drove the T-bob at the pithead. This has been sawn into two pieces, one of which is projecting up at an angle near the gear train.

The T-bob was constructed like an inverted letter T, on one arm of which were the pump driving rods, or spears, down the pit. On the reverse arm was a ballast bin holding some five and a half tons of ballast as a counterweight. Connected to the vertical arm, the stem of the T, was the horizontal connecting rod.

As the gear train driving these rods, etc., needed to be securely fixed to the cast iron base-plate, some form of clutch would have been necessary if winding was also to be carried out, as continuous rotation for pumping with the winding sheaves engaged would not have been possible because of the limited amount of rotation allowed by the chain drums. This clutch would probably have taken the form of a dog clutch, in which a sliding component on the shaft would have been moved to engage, by means of axial teeth, a similar fixed component on the winding drums when winding was to be carried out. In this case, the drums were designed for chains rather than ropes. Pumping during a winding cycle could be carried on because the rotation of the engine would then be immaterial.

This engine, notwithstanding its early build, was quite powerful, and as well as turning the winding drums, chains, etc. was also capable of pumping between nine and twelve thousand gallons of water per hour up to the watercourse which was 80 yards down the up-cast shaft. This water then flowed through a tunnel to near the Clog and Legging Colliery, then turned east, with the water exiting near the sawmills on the Albion Road and quite near the tunnel which starts near the Drill Hall in Pontnewynydd.

There is a school of thought which makes this engine the first on the site, doing duty as both pumping and winding engine. This is borne out by an inventory taken in 1852, when the pits were leased to a partnership, which, for historical interest, I have included at the end of the book (see Chapter 13).

In this inventory, only the beam engine is recorded, the vertical winder is not mentioned, but there are chain barrels (winding drums), two flat chains and a chain barrel shaft. There is also a large rectangular pit behind the smaller of the two gears; this still has the remains of several holding down bolts in slots in the walls. There is also a second wheel pit next to the flywheel pit. The purpose of this is unclear, unless it was for a brake wheel. There is, at the end of this pit, a steam pipe brought up from below ground level; again, the purpose is unclear.

Another curious point is the varying shapes of the plummer blocks used at the Glyn. Some of these have rounded ends, some have square ends, one has a square end with a buttress on it. There are about four different types, and some of these blocks on the vertical winder are also of this type. All the parallel motion links are now missing from both ends of the beam.

Another feature of this engine is the expansion joints on the steam pipes (see Fig 78). These allowed the pipes to expand without breakage when both ends were fixed rigidly to the valve chest, as here.

It was common practice then to use one engine for both pumping and sinking the shafts until any deposits – coal, ironstone, etc. – were proven, when a dedicated winding engine would be installed.

Figs 58 and 59 (opposite): Two views of lost motion on valve operating rods from below.

Fig 60: Close-up view of slot forged on valve operating shaft eccentric rod, partly lifted off the roller.

Fig 61: View of hand valve-operating lever, with engine exhaust pipe exiting through wall at left, apparently bypassing the condenser.

Fig 62: View of the engine cylinder from above. This engine has a stroke of 6ft and a 30in bore. The lower set of valve gear can be seen through the cast iron girder work, as can the valve-operating levers which are crossed over. Notice the two stiffening bands around the cylinder. This method of construction is virtually identical to that of the vertical winding engine cylinder. The dovetail method of securing the floor support girders – of cast iron – can also be seen here. The vertical rod, just seen on the right of the photograph, was the piston rod from the beam to the condenser air pump.

Fig 63: View of gear train bed-plate. The plate at the far end was where the chain winding drums were fitted. There would also have been some type of clutch – probably a dog clutch – to engage and disengage the winding drums. The brickwork now visible was part of a ventilation system and was not built until a later date. There are no mounting holes in the rear plate for bearing blocks etc. but the holding down bolts project high enough for another plate to be fitted on top of the present one. There are also several holding down bolts still in situ. Note the square holes in the flywheel rim. If the engine stopped 'on centre' these would have been used for barring the engine over centre for starting. Note also the different types of bearing blocks (plummer blocks). These wheels would have been quite noisy when working.

Fig 64: Anchor points for beam bearings to cast iron plate built into the wall. This method of securing these holding down bolts is identical to that used on the vertical winding engine; i.e. plate and cotter. The quality of the masonry is also evident here despite many years' exposure to the elements. It, too, is similar to that used on the vertical winder house.

Fig 65: Top arm of T-bob and connection to the horizontal connecting rod at pit-head. Looking carefully, one can just see the other arm of the T-bob between the vertical arm (bottom centre) and the brickwork, which carried the pump rods – spears – in the shaft. The main bearing is in the lower right corner of the photograph at the lower end of the casting.

Fig 66: This shows a view, from the second motion crank end, of the pushrod driving the top arm of the T-bob. This rod is of oak, reinforced with wrought iron straps. One can also see – under the rod – the top of the balance weight box attached to the near arm of the T-bob, with two long bolts as further support. This box contains $5^1/_2$ tons of ballast, the object of which is to balance the weight of the pump rods, or spears, in the pit shaft. These were connected to the far arm of the T-bob, which protruded into the up-cast shaft. These spears extended to the pit bottom and drove two separate types of pump; one was a lift pump, the other a force pump. The forcer, as the lower pump was called, forced the water up to a lobby in the shaft, from where the lift pump brought it to the level of the watercourse, 85 yards down from the shaft top. Also to be seen is a spare crank-web acting as extra ballast. It seems that the trench enclosing this pushrod may also have been used in the past as part of the ventilation system. There is also another steel plate in the left foreground, the purpose of which is not yet known.

Fig 67: Looking down onto the balance weight and end of T-bob. This balance weight weighed 5 1/2 tons, and acted as a counterbalance to the weight of the wood spears and pump rods in the pit shaft. These spears drove the pumps, one at pit bottom and one part-way up the shaft. The wood beam, top, was the horizontal connecting rod from the secondary crank on the beam engine second motion shaft to the vertical arm of the T-bob. Also evident is the repair in the near arm of the T-bob casting. The two round bars stretching from the ballast box were to help support the end of the box. Note also the spare crank web resting on the top as ballast. The bearings of this equipment carried a considerable weight: roughly 5 tons of spears and pump rods in the shaft, plus 5 1/2 tons of iron ballast. There is a possibility that these parts may still be buried in the brick-lined trench.

Fig 68: Flywheel and gear train.The flywheel is 17ft diameter, with the two gears approximately 4ft and 9ft diameter. Note the scale ruler on the larger wheel. The secondary crank is mounted behind the larger of the two wheels. Note again the missing bearing caps. The brickwork on the right was part of a ventilating system. Also note the bolts securing the spokes in the centre of the flywheel. All the square bolt heads are in line.

Fig 69: Lower set of valves, with exhaust valve nearest camera. The connection to the cylinder, the nozzle, can also be seen under the scale ruler. The pipe at the far side is the steam inlet pipe. The valves are Cornish equilibrium double-beat type, driven by a single eccentric.

Fig 70: View of main crank, bottom of connecting rod, and part of eccentric of beam pumping engine.

Fig 71: Flywheel and reduction gearing of beam pumping engine.

Fig 72: Flooring beams, now without the cast iron floor plates. Engine exhaust, and manual valve lever. The exhaust pipe seen here exits the engine house through the south wall, where there is a vertical pipe projecting from the ground, which appears to be connected to it. It seems that this method of exhausting the engine bypassed the condenser, and on examination, it can be seen that the water supply to the cold well has been blanked off. The vertical pipe on the right of the photo is mounted on a T-pipe, the other arm of which leads to the condenser top. The vertical lever in the foreground was the valve-operating lever, used when the single eccentric was disconnected for winding purposes. By reversing the steam supply to the engine, this lever could also have acted as a braking system.

Fig 73: Shows the vertical cylinder of the beam engine at the Glyn Pits. It also shows, on the extreme left of the photograph, a T-headed lever. This was used to control the engine – by operating the four Cornish double-beat valves – for winding coal, men, etc. while the valve gear was disengaged from the single eccentric on the first motion shaft. This was done by lifting the valve rod out of engagement with the driving rod, as described elsewhere. The chain and hook hanging down near the cylinder were from the small crab-winch, which is still at the mine, although now badly corroded. Yet this too is part of the history of the mine, and therefore should be preserved. This photo also shows the lagging, probably asbestos, on the exhaust pipe from the cylinder.

Fig 74: Beam pumping engine valve gear. The vertical pipe on the right was the exhaust to the condenser; note the expansion joint. The steam inlet is on the left. This is the bottom set of valves; there is an identical set for the top of the cylinder. All were worked by a single eccentric on the main, or first motion, shaft. Note also the small wooden wedges driven into the clevis joints – bottom centre – apparently to take up some of the wear.

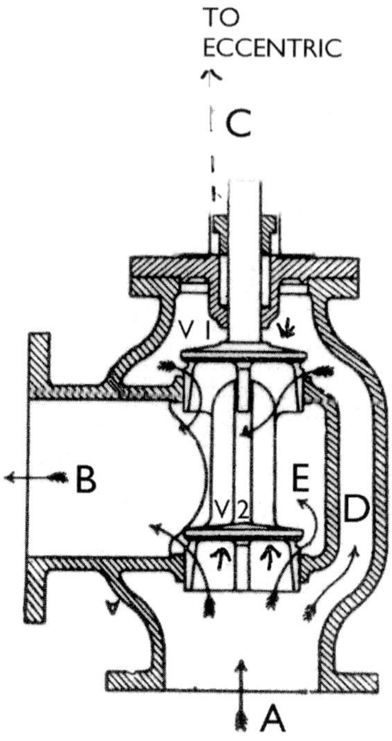


Fig 75: Drawing of a Cornish double-beat equilibrium valve. Steam enters at A and exits at B. Traversing the passage D, it presses down on valve V 1, attempting to close it; at the same time, it presses up on valve V 2, attempting to open it, so that the valve is in equilibrium, and can be worked without strain on either the operator, in the case of a hand-operated valve, or the driving linkage on a mechanically-operated valve. This valve is driven by the spindle C at the top, usually from an eccentric, but as the valve is balanced, it will drop closed, and only needs to be lifted open, so that in this type of valve, it is necessary to incorporate some lost motion in the linkage; this lost motion can be seen in the photographs of the beam engine valve gear at the Glyn Pits.

Fig 76: View of valve-operating levers, with rod to eccentric, centre right, and hand-operating lever, top. The end of the eccentric rod does not have a slot for lost motion. Instead, the rod is forged to a close fit around the roller on the bottom of the lever. This could then be lifted out of engagement to allow hand-operation of the valves, thus permitting the engine to be reversed for winding coal or men.

Fig 77: Valve-operating shaft, showing vertical valve rods, left, with a certain amount of lost motion. The throw of the main eccentric is in excess of that required by the lift of the valves, so extra motion is lost by these slots.

Fig 78: This shows the expansion joint of the engine exhaust pipe, centre, with lower valve gear behind. The steam inlet pipe is on the far side of the valve chest. Main cylinder base and bottom of cylinder are on the right of the picture. Note the valve-operating levers, which are crossed. The same rods drive both top and bottom valves, but the bottom right exhaust valve is on the same rod as the top left steam valve, so that when the top steam valve is open, the bottom exhaust valve is also open and conditions are the same for the other two valves.

Fig 79: A section of the vertical connecting rod on the beam engine. It shows the wrought iron reinforcing straps, and it also shows that the wooden part of the rod has been butt jointed in the past.

Fig 80: Bearing plummer blocks and centre of the beam at Glyn Pits. Note the strap over the bearing block. It is a built-up device bolted under one of the holding down bolts. It was probably for extra stiffness, as the forces at this point would have been considerable.

Fig 81: The cast spacer between the flitches (sides) of the beam. Note how the bolt heads are square with each other. This was commonplace years ago, and my father, when coach-building, always set the heads of the bolts, usually 1/4in bolts, square. Scale ruler is 2ft.

Fig 82: Shows the cylinder supports, centre, for beam engine cylinder, with the engine exhaust to the condenser lower right. The bracket bolted on at far left supports the valve driving gear.

Fig 83: Beam pumping engine parallel motion levers. Note the high quality of the workmanship on the turned rod, top left and top right. The top of the piston rod is just visible on the extreme right centre of the photograph, with two of the link rods leading to it. All these levers, including the parallel motion levers for the air pump, are now missing and the wooden anchor beams have rotted away, with just one anchor point still in place on the remains of one beam.

Fig 84: The beam from the small oiling platform. This platform was reached via a spiral staircase, now missing. Also worthy of note is the parallel motion. This particular set drove the condenser air pump piston. Note also the method of bearing adjustment. One wedge (cotter) at the top adjusts, through a turned spindle, both top and bottom sets of brasses. Just under the beam, and attached to the bottom brasses, is another spindle, with a bearing surface in the centre. This was the attachment for the air pump piston rod. This rod is still in place today, although again badly rusted. For some reason, the air pump and condenser ceased to be used many years ago, with the engine exhaust being taken out through a pipe at the side of the engine house, and the water supply cut off from the cold-well. The engine exhaust to the side of the house is still in position today.

Fig 85: Side view of beam, which is approximately 16ft in length. Note the method of securing the shaft in the centre boss with iron wedges. Note also the blacksmith-made strengthening bar over the beam pivot. There would have been quite considerable forces acting on these pivots. These bearings are anchored by square holding down bolts, which are secured through a cast iron plate built into the masonry on the east wall, also called the bob wall.

Fig 86: View of cold-well in the basement of the beam engine house. The cast iron cold-well tank is now in poor condition. Note the patch where the water supply from the small reservoir entered it. The water supply pipe has also been blanked off. Near the top of the cold-well, and entering through the square hole, can be seen the valve driving rod from the eccentric on the main, or first motion shaft. This rod appears to be made of wrought iron.

# 7
# The condenser

The idea for a separate condenser came from James Watt in 1765 but was not patented until January 1769. There were two main types; surface and jet. The condenser at the Glyn was of the surface type, although necessarily of an early design. This condenser is, with its associated air pump, fitted in a large cast iron tank, the cold-well, under the engine base, with a valve box containing a single flap valve between them. The air pump was driven, using Watt's parallel motion, from a link on the beam. The air pump piston is 3ft diameter, and has a number of square holes cast in it. These were originally covered with a rubber washer, or clack, the purpose of this is as follows: as exhaust steam from the engine enters the condenser, it is condensed into water, which contains some air. This water, or condensate, is then drawn by the air pump from the condenser, through the flap valve and into the air pump by the rising air pump piston. When the piston descends, the water, being prevented from returning to the condenser by the flap valve, is forced up through the holes in the air pump piston, lifting the rubber washer to do so. As the piston again starts to rise, the rubber washer, falling over the square holes, traps it over the piston, so the water is then lifted with it, to overflow into a channel which leads it to the hot-well, and thence for reuse in the boilers. Any air escapes to the atmosphere. In some later condensers, the square holes were cast on a slant. The water being forced through these angled holes caused the rubber clack to rotate slightly every stroke, thus evening out any wear.

The gains from a condenser are considerable. For example, the condensate returned to the boilers can be at a temperature as high as 110°c, which, when used as feed for a boiler, will take far less fuel to evaporate it, and it also causes less thermal shock to a hot boiler than feeding cold water into it. A further benefit is that this water being virtually distilled is already largely free of harmful salts, such as calcium sulphate. This reduces scaling in the boiler, again helping to increase efficiency, as a small amount of scale on the heating surface of a boiler will have a disproportionate effect on heat transfer.

However, the main benefit of a condenser is that it reduces the terminal pressure of the exhaust for, instead of the engine having to exhaust against atmospheric pressure, average 14.7 psi, which takes a considerable amount of energy, a condenser will reduce this to around four or five pounds, thus taking considerably less energy from the engine.

Water was always used for testing a boiler because if the test pressure proved too much for the boiler shell, causing it to rupture, no damage would be caused to the personnel or boiler surroundings, for as soon as the shell split, the pressure would fall, due to the fact that water is incompressible. The results with compressed air, however, would have been very different, and with steam, catastrophic, as the water used to generate the steam would flash into large volumes of steam the moment the pressure was released. This is the main reason why boiler explosions were so destructive, and why safety checks are so stringent.

When I served my own apprenticeship in the mid-nineteen-forties, boiler scaling and repair was a large and important part of our work. There was one incident which took place at a hospital where we regularly carried out boiler work. In this case, a low water level had caused the top of the furnace tube over the fire – which was not fitted with a fusible plug – to become red hot, allowing the steam pressure in the boiler to force it down almost to the top of the fire.

This was repaired using a large hydraulic jack – it weighed three hundredweight – and a separate high pressure pump. The jack was inserted in the tube with sufficient packing – in this case steel bars – to bring it to the correct height, then the top of the boiler tube was heated up from the inside, using oxyacetylene heaters, actually cutting lamps with very large nozzles! As soon as the tube was judged hot enough, pumping began, forcing the large swelling back into its place. This operation took many hours of hard – and extremely hot – work to finish to the satisfaction of the boiler inspector.

Finally, the boiler had to be tested, again using the high pressure water pump. To do this, every fitting had to be removed from the boiler shell, and the hole – mountings blanked off, then filling the boiler to the very brim, sealing it with another bolted flange, and raising the water pressure to around one and a half times the working pressure, then adding ten per cent!

Another article missing from the Glyn Pit (according to accounts) is the beautiful cast iron spiral staircase in the pump house. Part of this, the handrail at the top, can be seen on one of John Cornwell's excellent photographs, shown here (Fig 84), but I remember it clearly, and often wonder where it went.

During a recent visit to the Glyn, we found what appears to be a boiler. This is located on the north side of the vertical winder house, and is below the present ground level. I had already been told that there was a boiler buried there, but on closer examination, I do not believe it to be a boiler. I have several reasons for this: one, there are no flues in evidence, either in the shell of this artefact, or outside it; two, the position of the pipe entering this tank is in quite the wrong place for a boiler, being some three-quarters of the way up the side of the shell. There is no stop valve fitted, and in any case, steam drawn from this position would lead to a large amount of priming, a condition where water is carried over with the steam. This is potentially very dangerous, because being incompressible, any water trapped in a cylinder by a moving piston would burst the cylinder covers off, with potentially fatal results for anyone close to it. This pipe is led off in the direction of the original Cornish boilers.

I cannot say if this ever was a boiler, as over the years many modifications could have taken place, and of course, there is the presence of the stack at the rear of the engine house to account for. But we do know that there were two air compressors at the Glyn in later years, so whether this was a boiler modified as an air receiver or steam receiver to even out fluctuations in the – to us – very low steam pressure of fifty pounds, is a moot point.

There were many other uses of steam at the pithead, such as the beam pump, with a cylinder capacity of near 30 cubic feet, haulage engines, fan engine, air compressors, not to mention steam leakage from pipe joints and valves, which, from old photographs, was considerable. Low pressure steam, as here, requires considerably more volume to do its work than high pressure does, hence the size of the steam pipes at the Glyn, with the supply to the vertical winder, a low speed, single-cylinder engine, albeit a large one, without a condenser, being a nine-inch pipe. The exhaust from the engine is a massive twelve inches. A large exhaust makes it easier for the engine to expel the exhaust steam, itself a considerable duty.

When water is heated, boiling does not occur until a certain temperature is attained, depending on the pressure to which the water is subjected. Steam in contact with the water from which it was generated is termed saturated steam. In later years, particularly on rail locomotives and ships' boilers, superheated steam at several hundred degrees Fahrenheit was used. The higher the initial steam pressure, the higher the temperature, therefore the more work can be got out of it before condensation begins. Steam in some railway locomotives was being used at several hundred pounds pressure: in one case, Sir Nigel Gresley's high pressure four cylinder engine no. 10,000, built in 1929, used a boiler pressure of 450 psi.

In some modern power stations, steam is also still used at several hundred psi, while in the Benson boiler, water is contained in steel tubes, which are completely filled. At the critical pressure of 3,200 psi, the water turns directly to steam.

Fig 87: A view into the air pump cylinder. In service, the square holes were covered with a thick rubber washer. As the piston was driven down by the rod from the beam, so the water under it, in the pump cylinder, was forced up through these holes, lifting the rubber washer to do so. At the bottom of the stroke, the washer reseated itself, sealing the square holes, and preventing the condensed water – the condensate – returning to the cylinder. The beam then lifted the piston, together with the water, which then flowed out through a channel (see Fig 89) to the hot-well, where it was reused in the boilers. The piston rod, top of picture, was driven from the beam by Watt's parallel motion. In order to replace this rubber valve washer the piston rod could be detached by removing the four bolts and lifting the rod, complete with the hemispherical valve cover plate, (on which the scale ruler rests) clear. A sectional drawing of such a piston is given in Fig 88.

Fig 88: The adjacent drawing shows a sectional view of the air pump piston at the Glyn. The thick rubber washer or valve is marked A, the piston B and the hemispherical plate C. It can be seen that this plate allowed the rubber valve to rise only at the edges to admit water. This allowed the natural 'spring' of the rubber to reseat the valve at the end of the stroke. The piston rod marked E–D is simply a locating boss for the rubber valve. The bottom drawing shows the square valve holes in the piston. This particular drawing is actually a valve – one of four – from the piston of a later type of marine air pump, c.1940s. But except for the piston rod etc. it is virtually identical to the piston itself at the Glyn, as reference to the photograph above will show. This is the reason I have included it here. In some later air pumps, the square holes were slightly angled. This caused the water flow to rotate the rubber valve slightly at every stroke, thus giving even wear to the rubber.

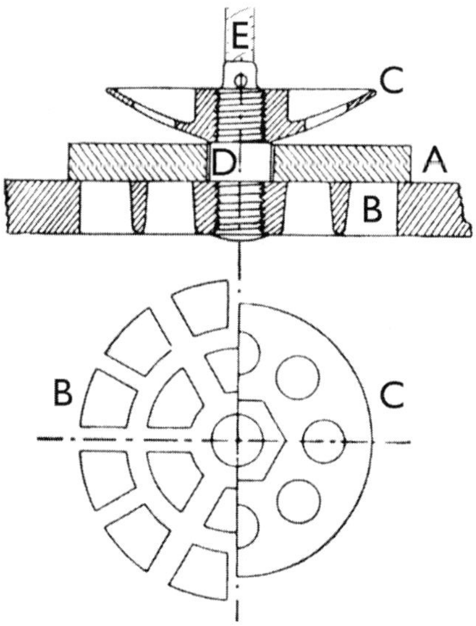

Fig 89: The condensate water outlet from the condenser air pump. In service, the channel from the pump and the exit channel were joined by a wooden trough. The vertical iron rod on the right of the picture is the piston rod, now long since disconnected from the overhead beam. This water was conveyed by this channel to the boiler hot-well, a large tank from which the boiler feed pumps drew their water. The original boiler house containing two Cornish type boilers at the Glyn was just outside this wall. Reusing water in this way had several benefits: it required less fuel in the boilers to reheat it, and since it had been heated in the boilers previously, it had already deposited the scale, and there was less thermal shock to the boiler shell.

Figs 90: Top of the control rod for snifer valve.

Figs 91: Snifer valve mounted on the condenser body.

Fig 92: Close-up view of a plug-type valve on the condenser body. This was known as a snifter valve, and could be adjusted by the engine man to admit just sufficient cold water from the cold-well surrounding the condenser for efficient condensation. The reason for using this type of valve in preference to a screw-down type was that just one quarter of a turn was required from open to closed. On the picture opposite will be seen a circular plate fixed to the engine cylinder base holding down bolt. This was not simply a support; the valve control handle was arranged so that the top of the plate could be marked with chalk etc. to give a setting mark for the valve-operating lever. There was a lever-type handle on this valve, not a wheel.

Fig 93: Condenser air pump. Square holes just visible under the top circular plate were to allow water to flow through when the piston was driven down by the beam. They were covered by a leather or rubber washer, which sealed the holes when the piston rose, lifting the condensed water, which then flowed out into the hot-well for reuse in the boilers. This water, being distilled, was free from harmful salts etc, thus helping to prevent scaling of the boiler. Also, being hot, it took less fuel to reheat the boilers.

Fig 94: The air pump is on the right, with the condenser in the centre. Between the condenser and air pump can be seen the non-return valve. This valve, by preventing water drawn out of the condenser by the air pump returning to the condenser, helps to maintain the condenser vacuum. In normal use, this cold-well was supplied with a constant stream of water from the reservoir at the rear of the pit yard, with used water overflowing on the opposite side of the cold-well. This then flowed into a drain which exited on the side of the tram-road approach to the Glyn, which can still be seen. The brass plug-valve on the condenser would have been opened by the engine man to admit some cold water to be sucked into the condenser to assist condensation, as unlike modern condensers, the surface area of this particular type was not enough for efficient condensation.

Fig 95: One of three cast iron steam pipe supports on the side of the beam engine house. Like bearing caps etc. used at the Glyn, this was probably a standard type of casting made at Neath Abbey Ironworks.

# 8
# VENTILATION

Mine ventilation is necessary to provide clean fresh air, to dilute and remove explosive and noxious gases from the mine, and, in some cases, to cool the mine workings.

In the early days ventilation, when used at all, was usually by furnace, making use of the fact that hot air, being lighter than cold air, rises. Later, as mines became deeper, other forms of ventilation became necessary, such as the Guibal fan, which was the first really effective ventilation fan, and was introduced to Britain in 1863 (at Elswick Coliery, Northumberland) and a development of this, the Walker fan. There was also the Waddle type fan. The Waddle was not like a normal fan; in the Waddle, the blading was enclosed between two discs and ran in the open air, with the mine waste air exiting around the edge, i.e. it was an open type. This too was introduced in 1863 (at Bonville's Court Colliery, Saundersfoot), but still fitted until comparatively recent years.

There was also the Nixon ventilator, which was a large square cast iron piston in a large square cylinder, with the piston running horizontally on two rails; this was given a reciprocating movement by a steam engine. It was apparently quite efficient, though not as efficient as a fan. It was introduced at Deep Duffryn Colliery, Mountain Ash, in 1859.

Another type used was the axial, similar to an aircraft's airscrew, but with several of them in line. Later, in the 1902, at Pelton Colliery, Durham, came the Scirocco type, with many small blades, running at high speed, which rapidly became the most popular type.

Ventilation at the Glyn in later years was by a small Scirocco type fan, (see Fig 97), about 3ft diameter, which was, at least in the early days, steam-driven through a leather or balata belt. This fan drew air through the mine via the down-cast shaft, through the mine workings (guided by air-doors often operated by small boys) and back via the up-cast shaft; from which the names up-cast and down-cast were taken; through a fan-drift – red brick at the Glyn – thence to the intake of the fan.

According to the Mine Abandonment plans obtained from the Coal Authority, the coal workings at the Glyn were quite extensive, and when one looks at the small size of the fan, it does not seem capable of really efficient ventilation of such large workings, and indeed this, added to the fact that the need for good ventilation was not properly appreciated or understood in the early days of mining, was a common cause of explosions in mines.

This remark applies even more forcibly to ventilation furnaces, which were not as efficient as even a small fan, not to mention the danger of a large furnace with its associated fire in a gaseous mine, itself the cause of several serious explosions. On at least one occasion, it was the cause of several men being gassed by the fumes from the furnace. It was customary to site the furnace in the bottom of the up-cast shaft, or a separate air shaft, and in order to minimise the risk of explosion by mine gases coming into contact with the furnace, mine air was drawn through a dumb drift, which started some 20 yards or so away from the furnace itself, then rose at about a 45 degree angle to enter the up-cast, or ventilation shaft, some 30 feet above the furnace top. Therefore the mine gases were not directly in contact with the fire, but were still caused to rise in the shaft by the heat from the furnace.

I could find no documentation on the ventilation system at the Glyn at all, but in one photograph, the up-cast shaft is not covered in to form an air seal, unless the pit was covered with flap doors, which were lifted by the ascending cage, and which was a fairly common practice, while on another photograph in my possession, the fan-drift appears to have a south-facing arch built into it. But the view on this picture is obstructed by trees. The present fan-drift, being of red brick construction, looks modern in comparison to other building works at the mine, which are for the most part of dressed stone.

The ventilation system in this, as in other mines, was controlled by air-doors. These could be made of wood or a material called brattice cloth or a mixture of both, and were, up until the early 1800s, usually controlled by young boys of seven or eight years of age, or perhaps by colliers who were no longer able to do a normal job through injury etc. These young lads would work the same hours as the miners, which could be up to twelve hours, much of which could be in total darkness, as they would have had to buy their own candles.

Fig 96 and 96a: Shutter cut in rear of fan casing to secure reverse air flow. The lever on the shutter – top centre – was to operate the shutter from outside the fan-drift.This shutter is fitted at the end of the left-hand airway in Fig 96a (below). The scale ruler is 2ft.

In a later section on the Quarry Level is included a photograph of a ventilation furnace and its associated chimney (see Fig 117-9). This is one of the very few remaining in Wales; in fact possibly only two.

It may not be generally known how one of the most important officials in mining – the fireman – became so-called. It was as follows: originally, part of the fireman's job was to burn off any gas which had accumulated in roof cavities. He did this by wrapping himself in sacking and crawling into the suspect area carrying a lighted candle on a long stick; he inserted this lighted candle into whatever he thought may be a gas-filled cavity, igniting any gas gathered there. A hazardous operation under the best conditions, but lethal in a poorly ventilated mine. How many explosions were caused in this way, without the true cause ever being known? A drawing of this activity may be seen in *Moston Colliery, Manchester, A Victorian Super-pit*, p.88.

> After the first day of January, Nineteen Hundred and Thirteen, or such later date as, in the view of the circumstances of the mine, may be fixed by the Inspector of the Division, there shall, in every mine in which a mechanical contrivance for ventilation is used, be provided and maintained in a condition to be put into immediate operation, adequate means for reversing the air current.
>
> Extract from the *Coal Mines Act, 1911*, Section 31 (Clause 3)

Fig 97: Remains of ventilating fan. The belt-driving wheel is about 1ft wide. The driving engine, originally steam, was behind and to the right of the photographer. What the original ventilation system was at the Glyn is not now clear, but it was probably by furnace-induced draught.

Fig 98: Underground haulage station, showing signal bell and signalling codes. Note also the gas-proof lighting fitting. This photograph was taken at Afan Argoed mining museum, near Port Talbot.

Fig 99: The route of the mineral railway between Glyn Pits, behind the photographer, and Blaendare in front. The Balance Pits were situated in the clump of bushes on the left of the track. The Balance House was in the trees on the right. The water pipe shown in Fig 100 was at the rear of the garden of the Balance House, shown on the map in Fig 3. There were several feeders, or ponds, nearby to feed the water-balance cages at this mine, the site of one pond being on the right, just beyond the gate. On the map can be seen a branch line behind the bushes to the left serving this colliery. It appears that later, after this mine closed, these feeders were used to supply the reservoir at Glyn Pits.

Fig 100: Water supply pipe to the Balance Pit. This pipe and its tunnel are at the rear of the site of Balance House. A local historian, Neil Waite, told me that the tunnel probably extends right through the mountain, known as 'The Darren', emerging above the firing range on Cwmlickey mountain. As may be seen, the tunnel is very small, just over 3ft wide and high, so digging it would have been exceptionally hard work. The reason for the wall blocking the far end is not known.

For more information on mine ventilation, see *The History and Development of Colliery Ventilation*, by Alan Hill, available from Ashbourne Editions, Ashbourne Hall, Cokayne Avenue, Ashbourne, Derbyshire, DE6 1EJ. The dates on p.89 are taken from this comprehensive book.

# 9
# THE BALANCE PITS

The Blaendare Balance Pits were situated about a quarter of a mile to the west of Blaendare, and about midway between Blaendare and the Old Glyn Pits. Sinking of the Balance Pit is thought to have started in 1832; today, all that can be seen is a depression on the side of the track. This track was the mineral railway from Blaendare, past the Balance Pit and Balance House, to the Old Glyn Pit, past the Glyn to the Quarry Level wagon shed, then to Glyntillery collieries Nos. 1 and 2, and finally to Hafodyrynys Colliery. From the Old Glyn Pit a line dropped down to the Pontypool Brickworks, then dropped further in the opposite direction first to the Blaendare locomotive shed, then past the site that was both Glyn Pits Nos. 3 and 4 Pits, and the mine we knew as the Clog and Legging or properly Baldwins Albion Road Colliery, then to the sidings alongside the main Pontypool to Crumlin main line. This line was also known as the Park Estate Line.

On a drawing of the Balance Pit dated 17 March 1833, there are shown two oval shafts, each of 25ft by 9ft, both built of stone 1ft thick, giving an overall size of 27ft by 11ft, and they were 88 yards deep. The two cast iron beams supporting the chain wheel were 7in wide at the end and 13in wide at the centre. It appears from the drawing to be around 19ft in length. The balance wheels over which the suspension chains were hung were 9ft 3in overall; the rim was 10in deep, and appeared to have been bolted in sections, and 9in wide, grooved to take the suspension chain links, which were from 1¼in steel, formed into links 6in long. This chain weighed about 44lb per yard, with a breaking strain reputed to be 64 tons. The cages were guided by guide rods, of 1½in diameter.

Also shown on the schedule are two brass valves and seats. These were probably fitted in the bottom of the water balance tanks, which were attached to the bottom of the pit cages, so that when each tank touched bottom, the valves lifted automatically to allow the water to drain out. The cage, with its loaded tram, would then be ready to be wound to the surface by the weight of the other descending cage with its full tank of water.

The wheel diameter to the centre of the chains was 8ft 6in. The load was 30 cwt. The water ballast tanks shown on the drawing are 18in deep, and appear to be of riveted iron construction, which was usual then, and held 35 cwt of water.

There were three reservoirs at the Balance Pit: the upper reservoir, holding 7,720 cubic yards of water when within 15 inches of the top; the middle reservoir, holding 1,420 cubic yards when within 18 inches of the top, and the lower, holding 1,800 cubic yards when within 12 inches of the top; a total of some 11,000 cubic yards of water.

I found it surprising that there was no provision for braking equipment shown on the drawing, as the balance wheel at the British, Talywaun (see Fig 127), was fitted with a hand-operated brake band. Yet, on the schedule for the Balance Pit for the early 1850s, there is shown clearly: a chain wheel, cog wheel, spur wheel, brake wheel, brake, standards, brasses, levers, valve boxes, discharging pipes, and air pipes.

An interesting feature is that there is what appears to be a small steam engine with a stroke of about $2^1/_2$in driving a gear wheel of around 2ft diameter. This gear wheel has the teeth shrouded (that is, the sides of the gear are brought up each side of the teeth) each side, and is meshed with a larger gear wheel about 8ft 6in diameter with a width of 4in and, for its width, an astonishingly large pitch tooth to tooth measurement of $3^1/_2$in, fixed to the chain wheel. This engine appears, from a hand lever shown on the drawing, to have had a hand-operated valve which looks on the drawing like a type of butterfly valve similar to a car throttle valve – a crank of 1ft $^3/_4$in stroke, which is very roughly fitted to a square on the shaft, and a very rudimentary set of slide bars. Also the sectional drawing of the cylinder shows a piston which does not look like a steam engine piston, while on the drawing there does not appear to be a steam supply pipe to this valve.

This raises the interesting question of why a steam engine was geared to a water balance wheel. The gears are in a fixed position, and so cannot be disengaged, which I think effectively rules out the theory of a stand-by engine in the event of a failure in the water supply to the mine. As there is no braking equipment on this wheel, I wondered whether this was indeed a small engine or, and I think this a better explanation, a mechanical–pneumatic retarding device for the wheel and cages, some form of which, given the weight of trams, cages, water ballast, etc. would be necessary. I have since been told by an engineer that it was indeed a hydraulic retarding device.

Fig 127 is a drawing of this balance wheel, dated 11 March 1833. It was found among the Hanbury Papers, at the Pontypool Park Estate Office, which I was able to study with the kind permission of Sir Richard Hanbury Tenison.

The cages are also very rudimentary, in so far as they seem to consist of a water tank base 18in deep, two $1^1/_2$in suspension bolts, angle tram rails, and a cross-beam 6in wide at the ends, and 7in at the centre.

Through the centre of each cross-beam is a ring-bolt. This appears to be about 2in thick, and has the suspension chain fixed to it, the whole being about 8ft high. The tram shown is of the parallel side type, with flangeless wheels. In this case, the flanges are on the rails, which are of an L-shape, with the flange on the inside.

On the side of the mineral railway opposite to the Balance Pit was the Balance House, and in the bank behind this site there is a small tunnel, roughly 3 or 4ft high, closed off with stonework, through which projects a large pipe with bolted flanges, (see Fig 100). What this pipe was for is not yet clear, but there is a strong probability it was a water supply to the Balance Pit. I was informed that the other end is to be seen near the Cwmlickey firing ranges, on the other side of the mountain, which was also a catchment area for the Upper Race reservoir.

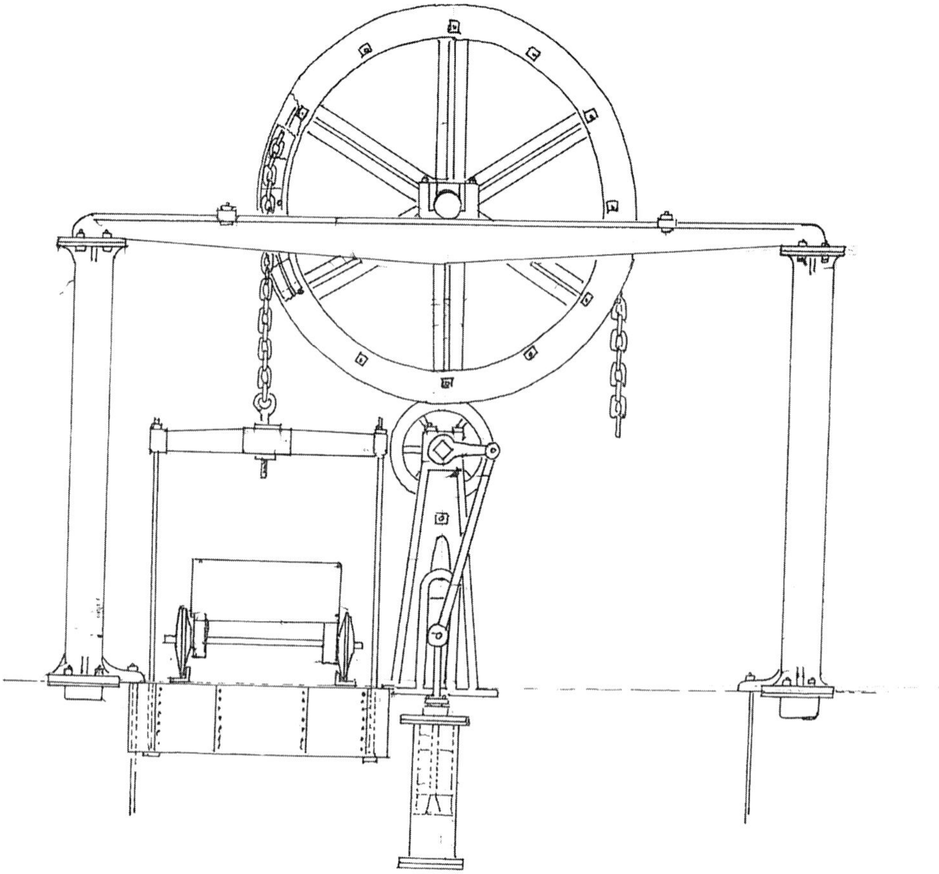

1 REV' OF WINDING DRUM = 26' 6' CHAIN
6 TURNS 159FT (SAY 200FT).
6 TURNS LARGE WHEEL = 33 TURNS SMALL WHEELS

ONE LIFT COULD TAKE 15 SECONDS. 30 SECONDS MORE USUAL.
IF WATER SUPPLY GOOD AT ONE WIND EVERY MINUTE, 720 WINDS IN 12 HOURS

Figs 101 and 102: Elevations of the balance wheel at Blaendare Balance Pits.

FROM DRAWING
"BALANCE PIT WINDER"
DATED 11$^{TH}$ MARCH 1833

MAIN DIMENSIONS

| | |
|---|---|
| WINDING DRUM | 9FT 3IN |
| CHAIN CENTRE'S | 8FT 6IN |
| PILLARS | 10FT LONG 1FT DIAM. |
| MAIN GIRDERS | 18FT OVERALL |
| MAIN PILLARS ARE | 6FT CENTRES |
| SMALL WHEEL | 2FT 4IN |
| CRANK CENTRES | 1FT $^{3}/_{4}$IN |
| HYDRAULIC BRAKE CYLINDER | 1FT DIAM. 2FT 6IN LONG |

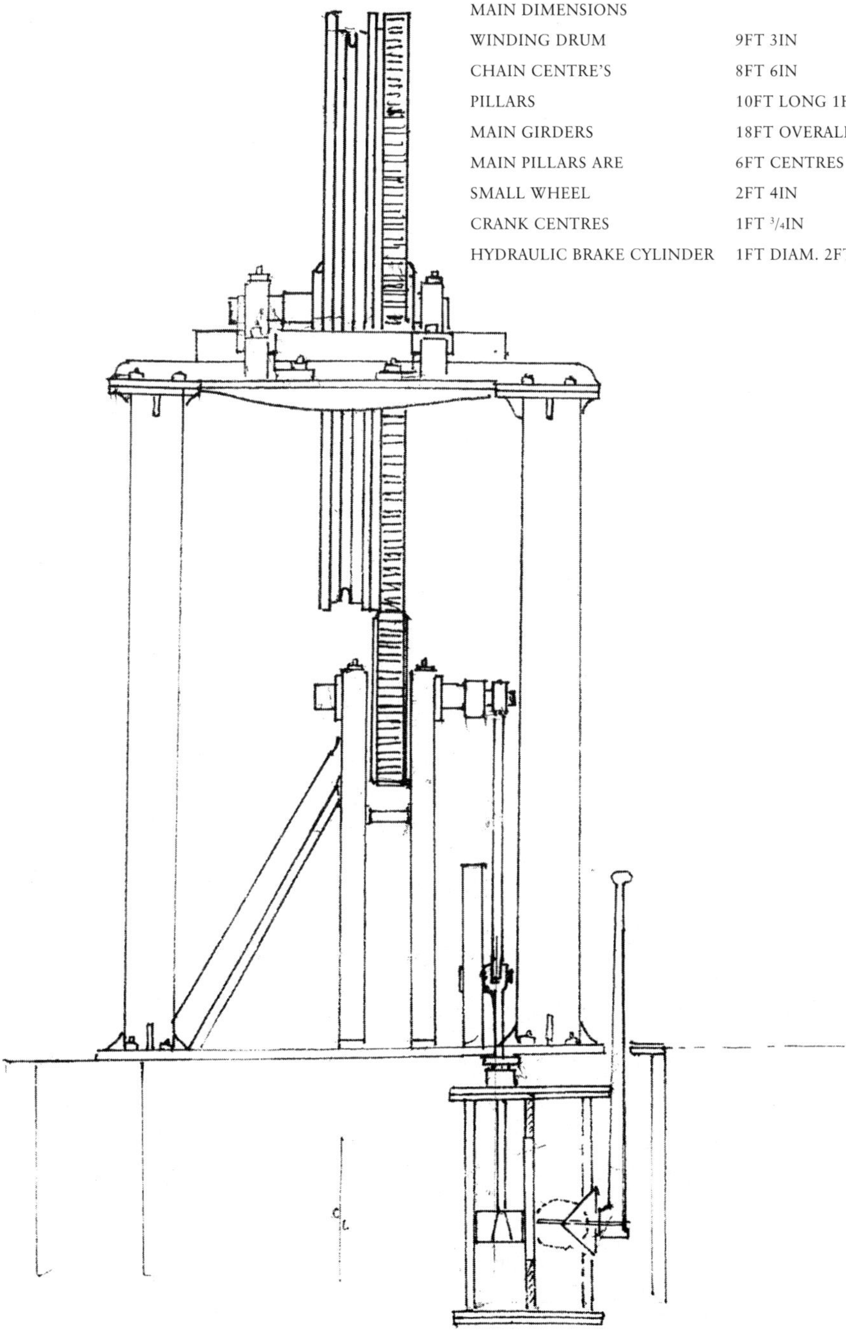

Fig 103: Water balance wheel at Cwmbergwrn, British, Talywaun, in Gwent. This balance wheel is rather different in construction from the one at Blaendare, although the working principle is the same. For example, it is fitted with a hand-operated brake band, where the Blaendare wheel is fitted with a hydraulic retarder. It is also considerably smaller and much less sophisticated. Unfortunately, this wheel has now been virtually destroyed after being dismantled by someone from Torfaen Museum some years ago. Large parts of it have since gone missing, presumably stolen for scrap, rendering it very incomplete.

# 10
# James Watt

James Watt, who at the time held the post of mathematical instrument maker at the University of Glasgow, was asked in 1764 to repair a model of Newcomen engine, which belonged to the Natural Philosophy Department. When he set the model to work, he was struck by the enormous loss of steam and efficiency because of the alternating heating and cooling of the cylinder.

He soon found many sources of heat loss. The cylinder itself was of brass, which was a good conductor of heat, then there was the cooling of the cylinder at every stroke to produce a vacuum, and also the power lost due to vapour pressure under the piston, the consequence of imperfect condensation. It occurred to Watt that a better vacuum might be obtained using a separate condenser, and after a period of intensive research, Watt was soon able to start his work of improving the steam engine. He soon saw that in order to reduce the working losses in the engine, it would be necessary to try to keep the cylinder always as hot as the steam that was used in it, as the great fluctuations of temperature during the piston strokes were the main cause of the inefficiency.

From the *Transactions of the Glasgow Archaeological Society*, 1859 we have James Watt's own words on the subject:

> I had gone to take a walk on a fine Sabbath afternoon. I had entered the Green by the gate at the foot of Charlotte Street, and had passed the old washing-house. I was thinking upon the engine at the time, and had gone as far as the Herd's house, when the idea came into my mind that, as steam was an elastic body, it would rush into a vacuum, and if a communication were made between the cylinder and an exhausted vessel, it would rush into it, and might be condensed without cooling the cylinder. I then saw that I must get rid of the condensed steam and injection water if I used a jet, as in Newcomon's [*sic*] engine. Two ways of doing this occurred to me: first, the water might be run off by a descending pipe, if an offlet could be got at a depth of 35 or 36 feet, and any air might be extracted by a small pump. The second was, to make the pump large enough to extract both water and air. I had not walked farther than the golf house when the whole thing was arranged in my mind.

As Watt told Professor Jardine, in referring to this invention:

> When analysed, the invention would not appear to be so great as it seemed to be. In the state in which I found the steam engine, it was no great effort of mind to observe that the quantity of fuel necessary to make it work would forever prevent its extensive utility. The next step in my progress was equally easy to inquire what was the cause

> of the great consumption of fuel. This too was readily suggested, viz. the waste of fuel which was necessary to bring the whole cylinder, piston, and adjacent parts from the coldness of water to the heat of steam, no fewer than from 15 to 20 times in a minute.

It was continuing this train of thought that led him to devise the separate condenser. On the Monday morning, Watt proceeded to make an experimental test of his invention, using for his steam cylinder a surgeon's large brass syringe, 4in diameter and 10in long. At each end was a pipe leading steam from the boiler, and fitted with a cock to act as a steam valve. A pipe also led from the top of the cylinder to the condenser, the syringe being inverted and the piston rod hanging down for convenience. The condenser was made from two pipes of thin plate, 10 or 12in long, and about $^{1}/_{6}$in in diameter, standing vertically with a connection at the top with a horizontal pipe of larger size, and fitted with a snifting valve. Another pipe about 1in diameter was connected to the condenser, and Watt fitted it with a piston, with a view to using it as an air pump. The whole was set in a cistern of cold water. The piston rod of the little steam cylinder was drilled from end to end to permit the water to be removed from the cylinder. This little model worked very satisfactorily, and the vacuum was such that the machine lifted a weight of 18 pounds hung on the piston rod.

Thus came about one of the greatest inventions of all time, and one which is still used to this day, and without which the mighty turbines in our power generating stations – oil, gas, and coal-fired – and also nuclear power stations, would not be very economical, if indeed they ran at all!

# 11
# A general view of the area surrounding the Old Glyn Pit

The approach to the Glyn Pit is now usually off the Pontypool / Crumlin new road, which replaces the original route. This was from the Crumlin road at Old Furnace, over the iron railway bridge, the approach road to which was also the road to Gypsy Level. The entrance to this level is still there and was driven under the lands of Gelly-Pistyll farm and then up the old inclined tram-road.

After crossing the bridge, one came to the Blaendare mineral railway. To the left of this was the Albion Road Colliery, owned by Alfred Baldwin & Co. Ltd, and known then as Baldwin's Level, or to us as the Clog and Legging. It is marked now by a large depression at the side of the route of the Blaendare mineral railway. This level was opened about the turn of the twentieth century on the site of the pits known as Glyn Pits Nos. 3 and 4. The Clog and Legging stopped coal production after the 1926 General Strike. It was then operated as a pumping station for some years by the Crumlin Valley Collieries Ltd until Nationalisation in 1947, when it passed to No. 6 Area as the Albion Slope Pumping Station. Adjacent to this level, and a little farther up the line, were Blaendare Company's engine shed, the engine driver being a Mr Yemm, and next was Upper Race or Pontypool Brickworks. To one's right after crossing the railway bridge was the electricity substation. This substation fed British Nylon Spinners at Mamhilad, as well as Tirpentwys, Blaenserchan, and Llanerch collieries, and the Varteg area.

The Old Glyn Pits abandonment map for the Rock or Black Vein (which was 6ft thick) shows that Glyn Pits Nos. 3 and 4 shafts were adjacent to the line of the tram-road, but slightly to one's left, after crossing the bridge. This would put them between the viewer and the Clog and Legging mine. On the same map the shaft pillar is shown for these two shafts. This is marked as a 'One hundred yards barrier'. It was usual to make the radius of the shaft pillar roughly equal to half the depth of the shafts. Also near here is Mynydd Maen Colliery. After climbing the tram-road, one is facing the Glyn Pits. About a quarter of a mile away to the west on one's right was the Glyn Quarry and Quarry Level, the site in the 1940s of the Blaendare Company's wagon repair sheds, where I first began my working life. Behind this, and at that time complete, was the Quarry Level, famous or infamous depending on one's point of view, during the 1926 miners' strike.

During the summer months, the route of the mineral railway from Hafodyrynys to Blaendare past the Quarry Level is now a pleasant walk, with the smell of the pine forest on one's right, and with the only sound that of the breeze and the hum of traffic on the new Crumlin road, which is usually barely audible. However, it was not always so.

During the 1926 miners' strike there was confrontation here and at other mines for many days, until on 3 September 1926, at the Quarry Level, a baton charge was made by the police. Some days later, on 24 September, upwards of 800 miners and ironworkers gathered at the Quarry Level, where there was also a police superintendent with some fifteen policemen. The colliery manager there then was a Mr Joshua Hopkins.

Another baton charge was provoked when one of the miners threw a fishplate (a piece of steel plate for joining rail lines together) from high up on the bank near the colliery screens, knocking Superintendent Spendlove off his white horse.

It was during this so-called Riot against the Person of His Majesty the King that Alderman Arthur Jenkins, who was then the miners' agent, was arrested, together with seven other men – William Coldrick, Deputy Miners' Agent; George Edwards, collier and member of Abersychan council; Bryn Thomas; William Keenan; David Gullick; Augustus Jenkins, a checkweigher, of Pontypool; and one Japanese man, a tinworker in Pontypool Tinworks. His name was Ichito Nishigawa.

Arthur Jenkins received a sentence of nine months in prison for his part in this so-called riot; he was the father of Roy Jenkins, MP, later Lord Jenkins of Hillhead. He was also one of the very few men who gained lasting respect for his actions that day, and the clock in Pontypool Park is dedicated to his memory.

These strikes were caused by low, one might say starvation, wages, token payments, by which men were paid with tokens which could only be spent in shops owned by the mine owner, often at an exorbitant profit, and worst of all lock-outs by the mine. It has left a legacy of bitterness among miners and others that is still with us today, culminating in the closure by Margaret Thatcher of most of Britain's mines, throwing thousands of men on the dole, and forcing Britain to import coal from abroad.

Just before these closures, I was visiting a friend in Cheltenham, and on having to wait for a large parade which was in support of the miners to pass the road junction, I asked a policeman why the people of Cheltenham were demonstrating for this reason, since there were no coalfields there. His reply came in one word: 'Dowty's.' Then I remembered that the Cheltenham firm of Dowty made, as well as aircraft parts, hydraulic mine supports or pit props. So the pique of this one woman hurt many other people also.

The remains of the Quarry Level screens can still be seen, as can the remains of a small cabin, built into the bank under the screens for the screen workers, and the route of the tram-road from the quarry, and later the Quarry Level, can still be seen. Water from the Quarry Level was used to supply the small reservoir at Glyn Pit, via a feeder around the side of the mountain. This feeder is still in place and can be easily traced, near the Quarry Level tram-road in the form of a large stone structure. Adjacent to this feeder were the stables, blacksmith's shop, etc. for the Quarry Level. Water was also taken from the Quarry Level to supply the reservoir at the Upper Race; these pipes of six-inch cast iron, which followed the mineral railway track to Blaendare and the Upper Race, were laid by the Pontypool Gas and Water Co and many of them can still be seen in place. They were connected to pipes from Glyntillery Colliery by a T-joint in a brick chamber, which is still in place today at the side of the mineral railway. Water from these collieries was conveyed to a large iron tank and pumping station near the Glyn Pit, as was water from the Glyn Pit itself, and it was pumped from there to the reservoir at the Race.

The Quarry Level is one of the very few, possibly just two small mines left today that were ventilated by furnace-induced draught, with the furnace still *in situ* (see Fig 117), although the chimney, also pictured, has now been demolished for safety reasons.

Also still on this site are two short lengths of blacksmith-made chain, roughly 30in long, which were part of the guying system of the ventilation stack. Each link of this chain was individually forged; consequently no two links are the same size. It is probable that these chains were forged by a Mr Bishop, who was a blacksmith at Quarry Level for many years. He later moved to Blaendare, where it is believed he worked with the blacksmith there, Mr Ted Roberts, and his striker, Mr Will Little, until near his eightieth year.

It is understood that water in the Glyn Pit has risen to the level of the main watercourse, which is 80 yards down from the top of the up-cast shaft, and it was from this position, in one report called a lobby, that Pontypool Water Company extracted their water.

Farther to the west down the track of the mineral railway, toward Hafodyrynys, and adjacent to the Upper Glyn Pond, was another small drift colliery, Glyntillery Nos. 1 and 2. In 1944, while working at the Blaendare wagon repair shed, my father and I were requested to go to Glyntillery No. 2 mine. It seems that the people of Hafodyrynys were being disturbed by a loud ringing noise from the main gear wheel of the haulage engine. Our job was to cut cheese-shaped sections of oak wood and bolt them between the spokes of this wheel, to deaden the sound. I am glad to say this solution worked.

This mine was also the setting for a small but funny contretemps, between the engine driver, Mr Yemm, and the mine manager, who was very conscious of his small stature. It was the engine driver's habit, when parking wagons at Glyntillery No. 2 colliery, to leave a break in the line of wagons for people to walk from the pit approach through to the pit yard. This particular morning he had not done so, causing the manager to call to him from the yard, about the same height as the locomotive footplate, 'I say, driver, you haven't left a gap for me to walk through.' Whereupon the driver, who was notoriously waspish, said, 'You are short enough to walk under the bloody buffers.' This episode ended with these two men, both in their early fifties, pulling faces at one another.

To the east of Glyn Pit, and again about a quarter of a mile away this time toward Blaendare, was the Balance Pit, now only marked by a small depression near the line. These pits were sunk to a depth of 88 yards in the 1830s, again at the direction of the Hanbury family, and were worked by water balance. On the same abandonment map, this pit is also shown as being connected to the Old Glyn Pit at shaft bottom.

The date given for the Balance Pit, i.e. the early 1830s, seems to me rather late. It was worked by a water balance, yet only some 10 or 15 years later, we see steam being employed for both winding and pumping at the nearby Glyn Pit. This was borne out by a friend, a mining surveyor, who advised that somewhere he had seen an earlier date than this for the Balance Pit, but could not confirm it. This date was in the late 1700s. In view of the earlier observations, and the later ones concerning the Aberdare Levels, 1797 would seem to be a far more likely date.

There was also a sad side to those days. For example; miners walked home from Glyntillery to the Upper Race, Blaendare, and Cwmynyscoy. Glyntillery was a notoriously wet mine, and even in the summer these men were soaked to the skin, but in the wintertime, when walking home into a bitter east or north-east wind, their clothes very often froze on them, so that they flapped like boards as they walked, and this for wages that would be considered derisory today.

The sad side of those days is also emphasized very well in some of the accident reports from the mid-18th to the turn of the 19th century and beyond. There are numerous reports of accidents where the remains of the victim were taken home in a sack. One can only try to imagine the dreadful effect on the man's family, whether parents or wife and children, when a sack containing the remains of their son or husband was dumped in their house.

Farther around the line was Blaendare Colliery. There was a Blaendare Colliery being worked by one John Wightman and Co Ltd from around 1865. The Blaendare Company was established in the latter half of the 19th century and working on Blaendare Slope by about 1873. This company was controlled by the Bythway family, who were also involved with other collieries and works in the area, notably those on Varteg hill. There were a number of levels or slopes and deep mines on the northern slopes of Mynydd Maen, the Cwm Glyn valley side. Some of these were: Wern Tillery (Quarry Level), Baldwin's Albion Colliery, and Glyntillery Nos. 1 & 2. On the south side of Mynydd Maen were Blaendare Colliery, Clay Slope, Aberdare Levels 1 & 2 (ironstone), Cwmlickey pits, Mountain Level, and Red Ash Level, with the Balance Pits (ironstone) on the south-eastern side. Just above Blaendare Colliery is Blaendare Farm, behind which and to the left was another entrance to the Quarry Level.

Just above Blaendare level are the remains of the Cwmlickey pits, with only the tops of the shafts still visible today. These pits were 98 yards in depth, and were wound by a high pressure engine which had a 20in cylinder by 5ft stroke. It was worked using an eccentric drive to a slide valve and was a beam engine, as evidenced by the fact that it also used parallel motion. Winding was by flat chains, four links in width. There was a single boiler to generate steam, 23ft long and 6ft diameter; it was made of half-inch plates.

The level known as Aberdare No. 1 opened c.1797, with No. 2 a few hundred yards away near an area known as Mountain Level. Aberdare No. 1 level is shown on the Ordnance Survey map as an ironstone level. It is understood that one of them, No. 1, runs out under the Upper Race reservoir to the area known as The Scourings, a name descriptive of the way ore was extracted from it using water. Strangely, it appears that the ventilation stack for this mine rises in the middle of the reservoir.

As one of the bolt boys at the wagon shed, it was part of my duty to go down with another of the boys to the springs near the tip to fetch water. We always allowed sufficient time for a short rest in Dick Williams's cabin on the tip, sitting by his stove and surrounded by old coats and sacks, and many were the times we dropped off to sleep in the warm cabin, to receive a telling-off from the foreman for being late with the tea.

One of the bolt boys there was in the habit of leaving the nuts too tight on the rethreaded bolts to move by hand and this was the cause of some complaint by those who had to use them. On one occasion, one of the men complained to him several times in one day. His answer was: 'I'm here to put the ........ things on, not get them off.' On some of these occasions, tempers would fray a little. This was also frequent when the tea bottles burst; many of the men at the wagon shed carried bottles of cold tea to work and it was our job, as bolt boys, to put the tea bottles to warm in front of the large fireplace. Some of the bottles had the screw-in type of stopper, some had corks, and if they were placed too close to the fire without slackening the stoppers, they would burst, leaving the unfortunate owner without a drink. This was another cause of much bad language!

Apart from warming the shed, this fire served another purpose; this was to reclaim metal brackets etc. from used timbers, because, being wartime, metal parts, including bolts, were scarce, hence the need for us bolt boys. It was also supposed to warm the shed, but with very large doors at each end, large enough to admit railway wagons, it was unable to cope, and I can personally remember the steel handle of the large stocks and dies sticking to my hands with the cold one morning. We had no toilets, no electric light – we had large cast iron duck lamps burning paraffin on an open wick – and apart from the spring, we had no drinking water either; but after a while we got used to it: we had to.

# 12
# MINE GASES

**Oxygen, chemical symbol $O_2$,** is both the most common and most necessary of the elements. About half the weight of the known rock formations is oxygen. Water is eight-ninths oxygen by weight, and air about one-fifth

**Nitrogen, chemical symbol N,** is the main constituent of the air. It has neither colour nor taste, and in its ordinary state is harmless and inactive. Nitrogen contains nearly 1% argon and with this it occupies 79% of the volume of the atmosphere. It is not poisonous, but does not support life or combustion. It is the main constituent of afterdamp and most blackdamps, and is also an important part of explosives.

**Hydrogen, chemical symbol H,** is also an element. It is the lightest and simplest substance known and has the lowest boiling point. It too is colourless, odourless, and tasteless. Its specific gravity is 0.07, making it only 7% the weight of air. It is harmless when breathed, and any ill effects would be due to shortage of oxygen. It burns readily in air, and if mixed in certain proportions, will explode violently. A mixture of two volumes of hydrogen and one of oxygen will explode on ignition, forming water and steam.

**Carbon Dioxide, chemical symbol $CO_{2,}$** is also known as carbonic acid gas, because it forms an acid in solution with water. It is very abundant in the enormous thickness of chalk and limestone rock formations, some of which have a carbon content of around 44%. It is always associated with fires and is an important constituent of the afterdamps. It is also a normal constituent of shot-firing, and therefore is an important consideration in extensive shot-firing operations. When mixed with the nitrogen of the air after an explosion, it is known as blackdamp.

**Carbon Monoxide, chemical symbol CO,** is also known as whitedamp, and is mostly associated with fires and explosions. It is the result of incomplete combustion of coal and other carbonaceous matter. It is odourless and colourless. It is slightly heavier than air, and burns with a flickering blue flame. It also forms an explosive mixture with air. It is extremely poisonous, and is easily absorbed by the bloodstream, forming a compound difficult to eradicate, and it therefore prevents the blood taking up the necessary oxygen from the air. The effects of carbon monoxide are cumulative, and the longer it is breathed, the more stable this compound becomes.

**Methane, chemical symbol $CH_4$,** a hydrocarbon, is given off by the coal measures, and by beds just above or just below the seams. It is usually stored under pressure, sometimes of the order of 400 psi, and blowers of gas can remove large blocks of coal from the face. It is lighter than air, non-poisonous, has neither taste or colour, and any effects of breathing it are usually due to impurities or oxygen shortage. It will burn quietly with a blue flame, but if mixed with between 5% and 15% of air, it will explode when ignited. Above 15% the gas will not explode, but the maximum explosive effect is between these limits, at about 9.4%. Below 5%, the gas will not burn, except as a cap on a safety lamp flame.

**Blackdamp,** also known as chokedamp because of the effect of the shortage of oxygen. It is not a definite gas, but a variable mixture of nitrogen and carbon dioxide.

**Afterdamp** is a mixture of gases usually caused by an explosion. The amount of coal dust contributing to the explosion also has an effect on the amount of carbon dioxide present in the afterdamp, which constitutes a great danger. There are also other gases, and the oxygen shortage adds to this danger. Firedamp explosions give carbon dioxide and steam, with the oxygen of the air combining with some of the carbon of the methane to form carbon dioxide, and the rest of the oxygen mixing with the hydrogen of the methane to form water. This leaves the nitrogen of the air, which mixes with the carbon dioxide to form afterdamp.

In a large mine explosion, there is insufficient air present for complete burning of the firedamp, and the amount of carbon dioxide would then increase in proportion to the ratio of firedamp to air. The shockwave preceding a mine explosion lifts large quantities of coal dust; this would burn to give large quantities of carbon monoxide, which would, as the gases of the explosion cool, pull in more oxygen, and form, with some of the gases from the burning coal dust, another explosive cloud, which in turn could be ignited by burning debris from the first explosion.

Mine explosions are extremely violent, and have been known to blow the hoist cages out of the mouths of the shafts.

A fact sometimes overlooked is that the nitrogen of the air is left unaltered by an explosion; it is left in a very large quantity, since it formed four-fifths of the air originally present; although not poisonous in itself, it is unbreathable, and therefore as deadly as blackdamp.

# SECTION

# 3

Fig 104: The entrance to the mouth of Aberdare No. 1 Level. This level, (ironstone) extends under the Race reservoir, in the middle of which is a ventilating shaft, and out toward the area known as The Scourings.

Fig 105: View inside Aberdare No. 1 Level. One of the tramlines can be seen lower centre.

Fig 106: View into the reservoir at the pithead, Old Glyn Pit. This was breached in January 1890 to allow the contents, some 760 cubic yards, to enter the pit by the down-cast shaft, to extinguish a fire that had been burning for some days. According to a contemporary report, water was still being poured into the mine a month later. This reservoir was fed partly from the Quarry Level. The width of the breach in the wall can be seen clearly here.

# 13
# INVENTORIES

These inventories of July 1852, for collieries and other works in the Blaendare area, including the Old Glyn Pit (Race Colliery and Cwm-y-Glyn pits), Cwmlickey Pits, Balance Pit, and Engine Pit, are included purely as a matter of historical interest.

## Cwmlickey Pits

One high-pressure engine, working by eccentric motion and slide valve-parallel motion and all of bright-work.
One engine house with stone walls, wooden roof, slated.
Three boarded floors and two wood stair-cases.
One boiler, with tubes made of half inch plates. 23ft long by 6ft diameter.
Two sets of cold water pipes, each 458ft long x 4in diameter.
One boiler-house built with bricks, iron roof, slated, with one brick stack.
One fly-wheel shaft, crank, spur, crown wheel, chain barrel, shafts and carriages.
Two pits, ninety eight yards deep.
Two flat chains for ditto, four links in width.
Two sets of pit carriages, crosses, etc.
Two pairs of pit frames.
Two top pulleys, two lands for rollers.
Landing and flooring plates, top of pits, 1,104ft, [by] $1^{1}/_{8}$[in] thick, weighing 21 tons, 8 hundredweights, 3 quarters.
Ditto for top of inclined plane 467ft by $1^{1}/_{8}$in thick.
One shed for horses, built with stone and slated roof, 12ft 6in by 11ft 6in within the walls.
One lodge for horses, built with stone 9ft by 7ft within the walls.
One cabin built with bricks with slated roof 7ft by 5ft 6in.
One inclined plane 320yds in length.
One chain for ditto, made of $^{11}/_{16}$ iron.
One shed for horses at the bottom of the inclined plane built with stone with pan-tile roof 9ft by 9ft.
One cabin, with stone roof, 9ft by 7ft.

## Cwm-y-Glyn Quarry, later site of Quarry Level, or properly Wern-Tillery

One crane.
One good chain and gearing.
One lodge, built with stones, wooden roof – slated, 11ft 6in by 9ft 4in within the walls.
One inclined plane, 180yds in length.
One brake wheel, two chain pulleys, thirty six rollers and boxes.

## Cwm-y-Glyn Pits (Also known as Race Colliery & Old Glyn Pits)

One engine house, built with good dressed stones with strong red deal roof.
One low pressure engine, 30in cylinder, 6ft stroke, double beat valves, working by eccentric motion.
Parallel motion, gearing, and bright-work.
Two chequered iron floors.
One wood floor.
Two chequered iron stair-cases.
One crab windlass.
One barometer.
One steam gauge [*sic*].
Two boilers with tubes made of $^{1}/_{2}$in plates, 26ft 6in long each 7ft diameter.
One boiler house, built with bricks, with iron roof, slated, and a stack built with dressed stones.
Cast iron pipes, 34yds of 6in for conveying water from reservoir to cistern.
Ditto 8yds [of] 4in for conveying water from cistern.
Ditto 36yds for discharging steam.
Ditto 3yds for conveying water in the reservoir.
One fly-wheel, shaft, crank, spur, crown wheel, chain barrels, chain barrel shaft.
Pump wheel and crank wheel.
One new wrought iron pump shaft.
One pit 190yds deep, 11ft diameter.
One pit, 190yds deep, 9ft diameter.
Two flat chains, four links in width.
Two pair of pit frames.
Two top pulleys.
Two pair of guide chains.
Two pit carriages.
One pair of wrought iron crosses.
Two pair wrought iron [undecipherable word].
219ft of receiving and flooring plates at the top and bottom of the pits, $1^{1}/_{8}$in thick, 4 tons, 5 hundredweights.
One large crab winch and 240yds of tar rope 6in radius.
One reservoir with an embankment supported by a strongly built wall containing 744 cubic yards of water when within 9in of the margin.
One lodge built with strong slated roof, 13ft by 9ft 6in within the walls.
Stable built with stones, slated roof, 21ft by 14ft within the walls.
One smiths shop, built with stones, slated roof, 17ft by 12ft within the walls, with one fire-place.

## Pumps

Diameter of working barrel 12in.
Height of lift 40yds, and diameter of forcer proportionate.
Height of forcer 60yds, fixed in a substantial manner and partitioned.
One red deal forcer, rod capped in the joints, strapped and well screwed together.
One lift rod, capped and screwed.
One 'T' Bob, with a box to hold 5 tons, 10 hundredweights of balance for the forcer rod.
Two brass buckets.
Four brass seats with leather clacks.
One lodge room for water at the bottom of the pumping pit.
One cast iron cistern and lodge room for water at the top of the lift.

## Balance Pits

The upper reservoir, with strong embankment containing 7,070 cubic yards of water when within 15in of the margin.
The middle reservoir, with strong embankment, containing 1,420 cubic yards of water when within 18in of the margin.
The lower reservoir, with strong embankment, containing 1,800 cubic yards of water when within 12in of the margin.
Cast iron pipes for conveying water to the pits, 3yds, 9in diameter.
Landing and flooring plates at the top and bottom of the pits, 1,666ft, 1in thick, 28 tons, 12 hundredweights, 3 quarters.
Two pits, each 88yds in depth.
One flat chain, three links in width, made of $^{11}/_{16}$ square iron.
One round balance chain, made of $1^{1}/_{4}$in [bolts?].
Two pair of guide rods, of $1^{1}/_{2}$in round iron with screw joints.
Iron railing by side of each pit.
Two pair of brass valves and seats.
Two tanks.
Two pit carriages.
Two crosses.
Chain wheel, cog wheel, spur wheel, brake wheel, brake, standards, brasses, joints, levers, valve boxes, discharging pipes and air pipes.
One shed for horses, built with stones, wooden roof, slated, 18ft by 9ft feet within the walls.
One lodge built with stones, 10ft by 8ft.
One cabin for Brakesman, 8ft by 3ft 4in, made with iron plates, wooden roof.

## Weighing machines

Philip Williams, capable of weighing 4 tons.
One set of weights.
Machine house, built with stones, slated roof, 9ft by 8ft within the walls.
George Hopkins machine, capable of weighing 4 tons.
One set of weights.
Machine house, 19ft by 8ft in good repair.
New machine, capable of weighing 4 tons.
One set of weights.
Machine house, built with stones, 10ft by 9ft within the walls.
Mark Tanner's machine, capable of weighing 4 tons.

One set of weights.
Machine house, 10ft by 8ft in good repair.
One powder store room, built with stones, slated roof 10ft by 9ft within the walls.
One small reservoir near Mine machine at Blaendare.
One small reservoir near Hopkins machine at Blaendare.

## Engine Pit

Two sets lifting pumps, diameter of working barrel: $11^1/_2$in.
Height of lift to the [launder?] 25yds.
Two brass buckets, with leather clacks.
Two iron pump rods 2 'T' bobs.
One double crank shaft.
One crank wheel.
One barrel shaft.
One chain barrel.
One cog wheel.
Framing over pit with two pulleys placed thereon.
One overshot waterwheel, 26ft 6in diameter, 6ft 8in wide.
One penstock, made of iron plates.
One slope, 47yds in length.
One slope, 62yds in length.
172yds of chain made of $^1/_2$in iron.
Two large pulleys.
Four small pulleys.
Twenty small rollers.
Six large rollers.
One shed for horses built with stones, slated roof, 12ft by 9in within the walls.
One lodge built with stones, slated roof, 12ft by 6ft within the walls.
Cwm-y-Glyn reservoir, with a very strong embankment, containing 754 cubic yards of water when even with the overflow.

N.B. The remainder of the surface of this land is included in the inventory attached to the second schedule.

## Blaendare Furnaces

One low pressure blast engine, steam cylinder $52^1/_2$in diameter, 8ft stroke.
Blast cylinder, 105in diameter, 8ft stroke.
Parallel motion, links, low gudgeon, plug rod, caps, steam cylinder lid, and all the gearing and bright-work.
Hollow stacks, Valves, Cistern air pump, and Condenser.
One engine house built with stones and strong roof stated, the first floor partly chequered plates and partly paving stones.
One stair-case of chequered plates. Second floor steam end of chequered plates.
30yds of 6in pipes conveying water from the 10in pipes to the cistern over the regulator, which is 8ft deep, 7ft long, and 5ft 6in, made of $1^1/_4$in iron.
34yds of 4in pipes conveying the waste water from the latter cistern to the one under the engine.
125yds of 6in pipes conveying water at the back of the furnaces to the cast house.
No. 4 furnace now in blast.

3½yds of 24in pipes.
One large blast box and valve.
34yds of 15in pipes around the tuyeres and throttle valve.
19yds of 2in pipes conveying water to No. 2 Tuyere.
33yds of 2½in pipes conveying water to No. 4 Tuyere.
67yds of 2½in pipes conveying water to the cast house.

[Author's Note: 'tuyeres' pronounced 'tweers', are the pipes or nozzles, usually of cast iron or of firebrick, that convey compressed air or 'blast' into the interior and charge of the blast furnace. Cast iron ones are usually water cooled. The bottom of a blast furnace was surrounded by a large pipe; this was called a 'Bustle Pipe'. The outer ends of the tuyeres were fitted into, and received air from, this pipe.]

Fig 107: Ironmaking, Blaenavon ironworks.This picture shows part of a small type of blast furnace, called a cupola. A cupola would normally melt pig iron and scrap in comparatively small amounts, the output being cast iron. This part is the bustle pipe, and it was supplied with compressed air, or blast, by a fan or, in earlier days, a blowing engine. It is from this pipe that the air was blown radially inwards to the bottom of the furnace – called the bosh – through tuyeres. The firebricks of the lining were made at Craig Ddu brickworks, at Cwm-Nant-Ddu – The Valley of the Black Stream.

Fig 108: Close-up of blast furnace tuyere, Blaenavon ironworks.

Fig 109: Tuyeres built into a wall at upper brickworks area, Blaenavon.

Fig 110: Baldwin's Albion Road Colliery, also known as the Clog and Legging, near Pontypool, in Gwent. This slope was driven under the mineral railway to Blaendare, the Park Estate Line into the bottom of Glyn Pits No. 1 and No. 2 shafts. It was on the site of the abandoned Glyn Pits No. 3 and No. 4, which were just off the picture to the right. It closed for coal production after the 1926 General Strike, and was used as a pumping station by Crumlin Valley Collieries until 1947, when it became known as the Albion Slope Pumping Station. The old Crumlin road is about 50 yards to the left of the camera, and the rails seen in the foreground are from the level known as Gypsy Level, so called because near here was a travellers' campsite many years ago. Gypsy Level was driven out under the part of the Crumlin road known as Gas Pitch and under the lands of Gellypistill farm, on the opposite mountain to the Glyn Pit. I was told many years ago by an old friend – Mr Bill Watkins – that Gas Pitch was so-called because the travelling people used to drive an iron bar into the ground, whereupon natural gas would issue through the hole from the workings underneath; they would then use this for cooking.

Fig 111: An underground photograph of Gypsy Level, now flooded.

Fig 112: Quarry Level, near Glyn Pits. From Quarry Level tram-road showing part of the stonework supporting the water feeder to Glyn Pit reservoir.

Fig 113: Quarry Level, view from top of water feeder to Glyn Pits.

Fig 114: The small cabin built under the screens for screen workers at Quarry Level, near Glyn Pits. The plank seat still remains in situ.

Fig 115: Quarry Level. Return airway, showing water main and part of flue for ventilation furnace.

Fig 116: Quarry Level (Wern Tillery). View of return airway. The pipe on the left is a six-inch water main to the Old Glyn Pit pumping station, then to the Upper Race Reservoir. The tram rails can also be seen here, as can the poor condition of the roof. The mouth of this level was about 50 yards away to the left, but there was also another entrance above Blaendare Farm, on the Upper Race. The ventilation furnace chimney was built over this airway, with the furnace itself to the left of the camera. There was also, in later years, a small Scirocco-type fan installed, which also drew air from the airway, in the roof just in front of the camera.

Fig 117: Ventilation furnace in the return airway, Quarry Level. The front of the fire is formed of railway lines. As can be seen, there is still fuel on the fire-bed.

Fig 118: Flues for the return airway of the ventilation furnace.

Fig 119: Ventilation furnace chimney at Quarry Level, base and lower part. This had been lined with fire-bricks. Now demolished for safety reasons.

Fig 120: The bottom of the brick-lined section of the ventilation furnace chimney Quarry Level. The bricks are of standard size, 9in by $4^1/_2$in.

Fig 121: Remains of firebox and boiler for a small horse-drawn engine. I remember at least two of these at the Quarry Level. There was also other equipment left there, including a chaff cutter for the horses, and numerous pumps etc. The boiler barrel was fitted to the far end of this firebox. The water space can be seen, as can the firebox stays – the rivets holding inner and outer plates together. The firebox door is also double, to protect it from internal heat. These were called agricultural engines because of the use on farms for threshing corn, etc.

Fig 122: A group of surface workers at Glyntillery Colliery, near Pontypool, in Gwent. This mine, one of the wettest in the district, is about half a mile west of Glyn Pits. The reason for including this picture is the tram on the left of it. The end door of the tram can be seen clearly, and so can the gag. This is the piece of steel bar dragging behind the tram, between the rails. This was hooked on to the coupling of the last tram before the journey was hauled out of the mine. In the event of a tram, or perhaps the whole journey, breaking away because of a broken coupling or haulage rope – a not uncommon occurrence in the early days of the twentieth century – this device, the gag, would dig into the sleepers or the ground between the rails, and derail the runaway trams before much damage could be done. The effect of a journey, or even just one tram weighing over a ton when full, breaking free in a steeply sloping mine can be easily imagined. Note also how the sides of the tram have been built up to increase the capacity.

Fig 123: Remains of sluice gate above the area known as The Scourings, Upper Race. To the left of the gate is the small reservoir or pond, which was controlled by this gate; on release, the water exposed minerals for exploitation. This method of mining is more properly known as hushing. This small gate – scale is given by the film case – was probably lifted by a hand-operated screw.

Fig 124: Ironstone outcrop, Blaenavon, Gwent, with small veins of coal over it.

Fig 125: Bell Pits. These were sunk by hand to a few feet, typically 15 or 20ft, then expanded to a bell shape. They would be abandoned when it was judged the pit was unsafe. These pits have fallen in over many years. It was said that 'Another pit would be dug at the distance a man could throw a shovel of earth'.

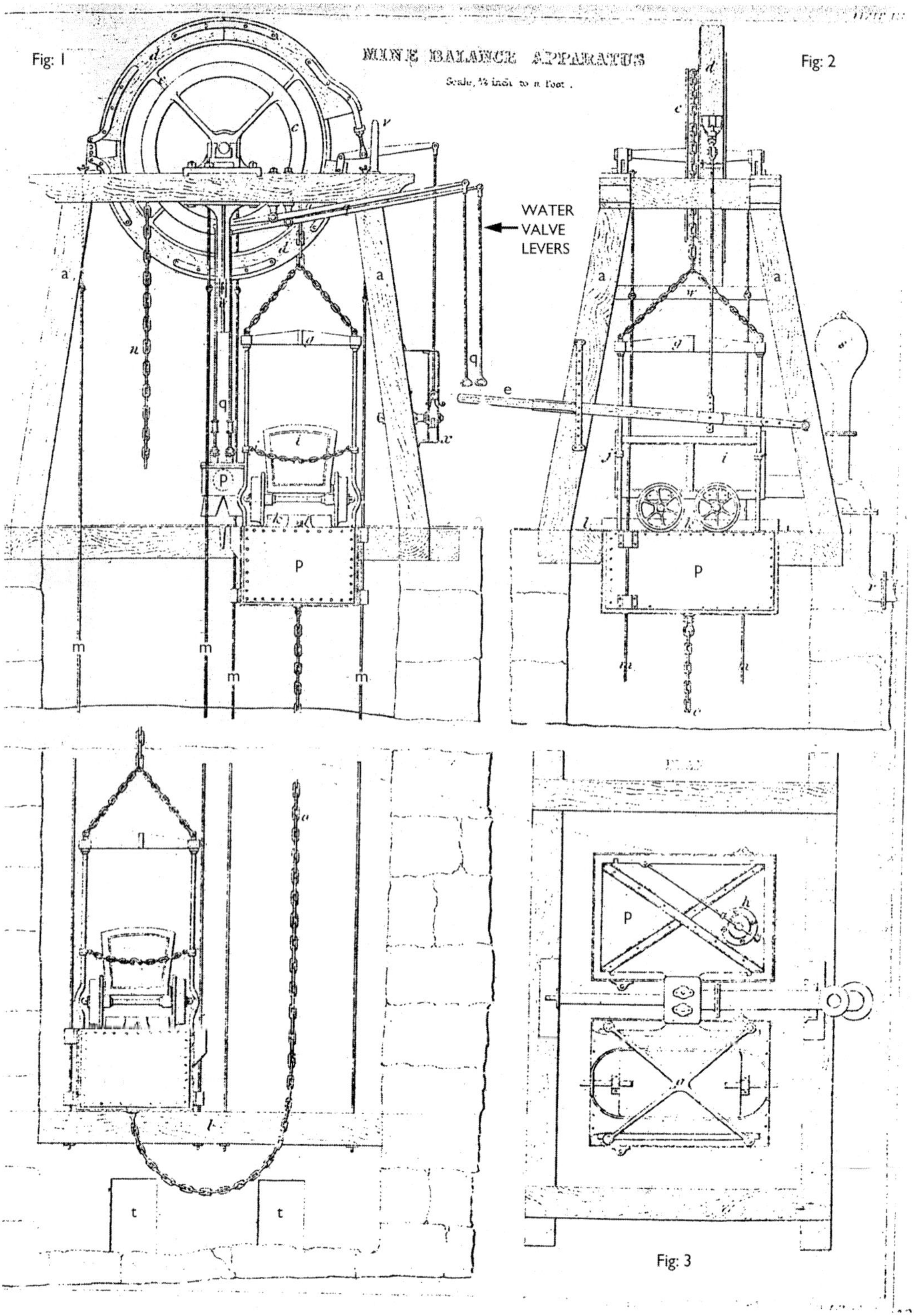
MINE BALANCE APPARATUS
Scale, ½ inch to a Foot.
Fig: 1
Fig: 2
Fig: 3
WATER
VALVE
LEVERS
a
a
a
a
e
m
m
m
m
P
P
P
P
q
q
t
t

Fig 126: (Opposite) An engineering drawing of one of the water balance wheels at Cwm Bergwm, Talywaun. This drawing refers to the upper of the two pits, and unlike the lower wheel, the main frame is of wood.

On the drawing marked Fig. 1, the two water valves, marked 'P', are set directly under the wheel bearings, connected to a pair of levers 'q' terminating in hand grips. These levers controlled the water supply to the balance tanks under the tram carriage. On the drawing, Fig. 2, the flanged water supply main is visible on the right, entering from below ground level. On the same pipe, just over the top bend, can be seen an air receiver, which is the device shaped like a bulb. The purpose of this was to minimise a condition known as 'Water Hammer'.

This phenomenon is caused by a sudden stoppage of rapidly moving water in a pipe, as by the sudden closing of a valve. It can be extremely destructive, even bursting pipes. However, if an air receiver is fitted the moving water is diverted into it, bringing the water to a standstill gently and compressing the air in the receiver, thus dissipating its initial energy. Water hammer can also occur in domestic pipework, resulting in a knocking sound from the pipes.

A tram can be seen in the carriage, secured from rolling out by a short chain across the entrance, while under the tram is a water tank. This tank usually held around two tons of water, which was allowed to flow to waste at pit bottom, and this is the reason why the depth of this type of pit is always limited by the valley floor.

A chain is attached to the bottom of both cages. This is to balance the weight of the hoisting chain, as one cage is lowered down the shaft, the weight of chain is considerable, but by using a balance chain as shown, this weight is cancelled out.

On the side view of the wheel can be seen the brake mechanism. This is hand operated by the lever shown in Fig. 2, and was, as can be seen, built of wood blocks tied together with wrought iron straps. The operator was able to lock the brake lever 'e' in the 'on' position by means of a pin through the holes in the bracket fixed to the left 'a' frame.

The drawing bottom right is of the cages, with the top drawing showing a plan of the interior of the tank bottom, with water release valve and operating lever.

The bottom drawing is of the other cage, and shows just the two semi-circular guides for the tram wheels fitted to the floor of the cage. It was not fitted with rails. It also shows the water supply pipe and valve tops between the cages, as well as the supply pipe on the right. On this drawing, there is a 'Bridle', i.e. four suspension chains, fitted to the cage tops. This was not normal practice for a water balance wheel. Water balances were usually fitted with a 'Cross' from which the cage was suspended by means of wrought iron bars, with the hoist chain fitted to the centre of the cross. This cross is shown fitted to these cages, and is mentioned in the accompanying inventory for Blaendare. The hoisting chain could be either of the flat link variety – similar to a very large bicycle chain – or of normal links.

The two rectangles marked 't' at the bottom of Fig: 1 were channels into which the tanks were drained. The guide bars for the cages can also be seen, marked 'm'. On this balance wheel they ran in guides inside the water tank. Although the working principle was the same for all balance wheels, there was no fixed formula for them. Design detail depended on where they were made, and the engineer in charge.

Fig 127: Water balance wheel at the British, Talywaun, near Pontypool. This is the lower of the two balance pits, with the remains of the other pit behind the camera and to the right. Nothing remains of the machinery at this wheel. This photograph shows the appalling condition of the balance wheel, one of very few left. Many of the parts have been stolen, including all the brake gear, wheel bearings, pipework, etc. and even one of the main cast iron A-frames which comprised the ends of the framework. These weigh roughly a ton each. The wheel can be seen lying on the ground to the right of the water bucket, with the rest of the mainframe scattered about the ground close by. The remoteness of the site is obvious. Although this wheel has nothing to do with the Glyn Pit and Blaedare area, I have included these photographs because they show the working principle of this type of machinery, and as far as I am aware, there are no photographs of the Blaendare wheel extant. The waste tips on the right are from the now disused Cwm Bergwm Colliery.

Fig 128: This heavy cast iron T-pipe, lying near the balance wheel, was for the bucket-filling apparatus. As can be seen, the two near pipes – one of which has been broken off – have a tapered shape. The water inlet to this pipe was up through the centre pipe (under the 2ft scale ruler) while the two top pipes, which have the flanges broken off, were where the tops of the two valves were fitted. These were operated by the hand levers shown on the accompanying drawing, Fig 126. Sadly, these unique valves, now irreplaceable, have also been stolen for scrap. In service, this would have been fixed in a vertical position, with the tapers looking down the top of the pit shaft, and when the valves were opened, in this case lifted, these would have fed water into the buckets by means of two corresponding water chutes on the buckets.

Fig 129: This shows the bottom of the water container. The centre eye-bolt would have had a chain attached to it which would also have been attached to the other water container. This hung down the pit under the container, and was to balance the weight of the hoisting chain hanging down the pit. The hoisting chain was of flat links, usually about three or four links in width, while the balance chain would have been of normal type links, as it was simply used as a balance weight. The bottom of the water outlet valve seating can also be seen, as can the two openings – right and left – in which the guide bars ran. These were to keep the cage assembly running true in the shaft. This too was identical to the Blaendare balance wheel.

Fig 130: This shows how the guide tubes were fixed inside the water tank, and, left, one of four supporting rods attached to the lifting cross. These are 2in square tapering to 1¹/₄in round. Scale ruler is 2ft long.

Fig 131: Water balance wheel at British, Talywaun. View inside the water tank. On the left can be seen one of the tubular guides, while at the top is the seating for the water outlet valve and the valve-operating lever. When the cage arrived at pit bottom, this valve was opened by pulling a chain attached to the end of the lever (lowest in the picture). This then lifted the valve off its seat, allowing the water to drain out. This waste water then ran into an adit or tunnel at the pit bottom, which conveyed the water to the nearest brook or drain. This method of winding limited the depth of the shaft to the lowest drain point in the valley, in order not to have to pump the water out of the mine again. As will be seen, the tank is of riveted iron construction. The iron bar over the valve lever was a guard to prevent the valve being lifted completely off its seating.

Fig 132: Water balance wheel at British, Talywaun. This view shows the supporting cross with the suspension link under the scale ruler. The hole in this link is very small considering the weight it had to carry; it is just $2^1/_2$in long by $^3/_4$in wide. The water container is $77^1/_2$in diameter by 28in deep, giving a volume of 76 cubic feet or 476 gallons, or over two tons when full. It was probable that the weight of water was nearer $1^3/_4$ tons or so, since the bucket would not have been filled to the brim. The two pipes seen each side of the bucket were for the guide rods, which guided the bucket and cage in the shaft.

Fig 133: Water balance wheel at British Talywaun, one of two A-frames. These frames were the end supports for the balance wheel and its associated mechanics. Sadly, one of these has been stolen, and as they weigh in excess of a ton each, heavy transport would have been needed to move it. There were originally two of these, and two other centre supports which were fitted under the wheel bearings. These bearings were mounted on large timber frames. This wheel was different in construction to the wheel at Blaendare, which was entirely of cast iron and was used for mining ironstone. (Scale ruler 2ft long.)

Fig 134: The rim of the balance wheel with the brake band track at the top, and the chain track below. The hoisting chain used on these wheels was a flat chain, usually three or four links wide, built up of flat plates similar to a bicycle chain without the rollers. This configuration explains the grooved bottom of the track. The brake was of the band type with wood brake blocks, and was hand operated. It would have been held in the 'on' position while the tram was loaded and the balance tank was filled, and at the same time, the full tram was being loaded into the cage at pit bottom and the bottom tank emptied of water. On releasing the brake, the weight of water in the tank, plus the tram, caused the cage to descend into the shaft, lifting the full cage – with its empty tank – up the shaft. A complete winding cycle would have taken around two or three minutes. The site of the water supply reservoirs for these two pits is unknown.

Fig 135: The brick-lined pit for the wheel in the accompanying photographs.

Fig 136: This shows the other, stone-lined pit. This one is closer to the Cwm Bergwm Colliery.

# 14
# THE DAVY SAFETY LAMP

I have included this chapter because the use of an open flame, in this case a candle, played such a large part in the social history of the Old Glyn Pits by being the direct cause of a severe underground fire, which claimed the lives of five men and nine horses.

Sir Humphry Davy (1778–1829), was an English chemist, also a pioneer of electrochemistry and the inventor, in 1815, of the miner's safety lamp. After an indifferent schooling and an apprenticeship with a surgeon, he discovered nitrous oxide (laughing gas). As a result of this discovery he was invited to join the Royal Institution in London. He was also the discoverer of the elements sodium, potassium, magnesium, calcium, strontium and barium. Davy identified and named the element chlorine after he had demonstrated that, contrary to Lavoisier, oxygen was not a constituent of acids, and demonstrated that diamond was a form of carbon. However, it has been said that his greatest invention was his appointment of Michael Faraday – who, arguably, had one of the most efficient intuitions in the scientific establishment – as his assistant in 1813.

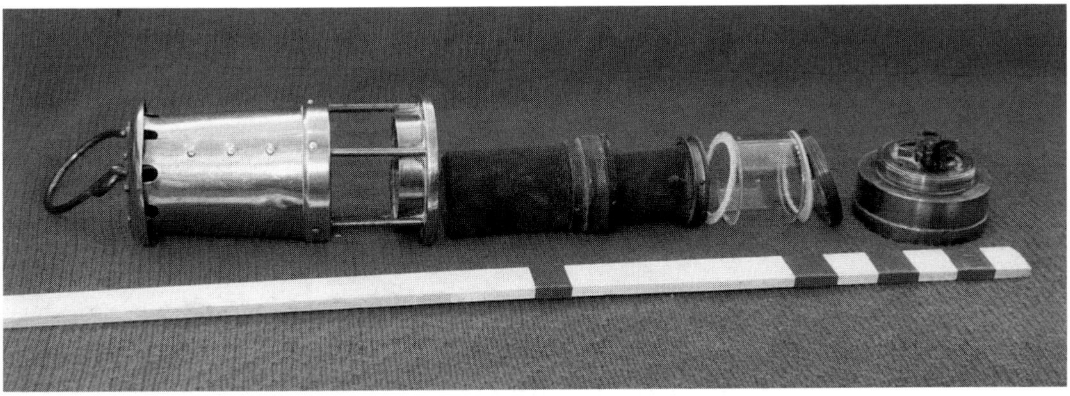

Fig 137: Exploded view of a miner's lamp. This lamp is known as a Garforth and was used chiefly by colliery overmen and firemen for gas testing. It is similar to the normal lamp, except it is slightly smaller. From the right are shown: the oil reservoir with, in this case, a circular wick which held sufficient oil for about nine hours' burning. Next is a securing ring. This held the glass, with its two asbestos washers, and the two gauzes together inside the bonnet of the lamp, giving a gas-tight seal. On the top of the bonnet is the suspending hook. This lamp has a flint lighting device, as for safety purposes the lamp would normally be locked, and could not be opened without specialized equipment. In this case the lock is magnetic, and the bench for servicing these lamps had a large electromagnet incorporated in it.

The safety lamp, as invented by Davy, relied on the fact that by surrounding the lamp flame with copper wire gauze which conducted any heat away, having a cooling effect on flame or hot gases, it prevented the heat of the flame or hot gases of a lamp igniting firedamp outside the lamp in a mine atmosphere. In the early type of lamp there was no glass, so that the light from the oil flame was considerably reduced. In later types of lamp, the gauze cylinder was fitted only in the top part of the lamp body, with a thick glass surrounding the flame. Both gauze and glass were fitted with a gas-tight seal. For safety reasons, the body of the lamp was arranged so that it could not be opened other than with the proper apparatus, and in a safe place, usually using either an electromagnetic locking device built into the workbench or a small padlock to secure the base from unscrewing. The safety lamp was, and still is, used for gas detection underground.

Firedamp, or marsh gas, being chiefly hydrogen, is therefore lighter than air, and so collects in cavities in the roof of a mine, so that by holding the lamp, which draws its combustion air from the top of the lamp, up near, or into, a cavity, the gas would be drawn into the lamp and cause a cap on the flame. The size of this blue cap over the lamp flame would indicate, to a practised eye, the amount of gas present in that cavity.

The early Davy lamp was not safe in an explosive mixture of firedamp and air if the air current was moving at about 5 or 6 feet per second. Modern lamps are protected by a bonnet or shield, and are arranged so that it is impossible for the flame to pass through the lamp.

With the original Davy lamp, it was possible for the flame to be blown against the gauze, making the gauze hot enough to pass a flame. Similarly, the pressure from a blown-out shot, even at some distance, was also sometimes sufficient to force the flame through the gauze, and if an explosive mixture of gas/air was present around the lamp at the time, an explosion could follow. A very similar type of lamp was invented almost simultaneously by George Stephenson. Safety lamps were in use at the Old Glyn alongside candles.

# 15
# YOUNG PEOPLE EMPLOYED IN MINING AND IRONMAKING

Children and young people were employed – perhaps abused would be a better term – by local industries in the 17th and 18th centuries. Detailed information about conditions is to be found in a report written on or about 5 May 1841, on the Pontypool District, to Her Majesty's Children's Employment Commission. It was written by a Mr Rhys William Jones, with interviews of the staff and children at various workplaces in and around Pontypool.

> Gentlemen; in the Parish of Trevethin, in which the town of Pontypool is situated, there are several large and important ironworks, tinworks, and collieries, in which, and in the mines connected with them, there are great numbers of children and young persons employed.

Mr Jones then goes on to say:

> It is very difficult to obtain any correct information of the aggregate number employed, as the returns from all the works have not yet been received, but it may be estimated, including the tinworks and collieries in the adjoining parishes between Pontypool and Caerleon, at from 1,500 to 2,000.

Following this, Mr Jones goes on:

> The youngest age at which children in the neighbourhood are employed in the works is between six and seven years. The fathers (if colliers or miners in particular) frequently take their sons, and sometimes their daughters, at this age, to carry their tools to be sharpened and repaired, or to open and shut doors to direct the air-courses in the collieries, or to do some other light description of work that may assist them in their labour, and increase their weekly earnings.
>
> The number, however, taken to work at this age is not great, probably not more than one or two children may be found on the establishment of each of the works.

I was fairly well informed of some of the working practices then regarded as normal, but I was quite unprepared for some of the cases recorded here, one such being that of a small boy called William Brockwier.

Little William worked in the Abersychan Coal and Mine Works, which belonged then to the British Iron Company's works. William's statement reads as follows:

> I shall be seven years old on the first of August. I keep this door. I have been here ten months; I do it very well; I open it every time the haulier comes, I don't know how many times in a day, one horse has been through twenty-five times; there are two horses at work. I come down the pit by myself, sometimes in the carriage they put me into, I lie down in it. I like to come to work better than stay home. I have no mother, I have a father, and two sisters, they are younger than me. I don't know what I get, I go to the pay sometimes, but my father gets my money. No one beats me here, my father beats me sometimes at home; I always have a candle, I have four candles in a day. I go to Sunday school, but can't read.

This child started work at just over SIX YEARS OLD. and was a door-boy underground. This door, as stated earlier, controlled part of the ventilation in the mine, so that here we have a situation where such an important thing as mine ventilation was left in the hands of a child of six or seven years, a child who worked the same hours as the adults in the mine; a shift of between ten and twelve hours. Most of this time he was alone in the mine, with just his candle for company. We must also remember that it was customary for the workman to buy his own candles, and, in the case of miners, their own gunpowder for blasting, too.

In this context, we should also remember that the massive mine explosions at both Llanerch (1890: 176 dead) and Senghennydd (1913: 439 dead) were caused by a build-up of firedamp, because of poor or non-existent ventilation. These were just two of hundreds of explosions which claimed thousands of lives.

On the subject of holidays at the British Iron Company at Abersychan, a Mr William Wood, aged 48, manager of the works, said:

> We do not profess to keep any holidays, excepting Christmas Day, but all our works except the blast furnaces are stopped, in consequence of the men being drunk, mostly two or three days at Christmas, and also on Easter Monday, which is generally a good day for club feasts.

Mr Wood goes on:

> All precautions are taken to prevent accidents that are practicable. Ventilation is particularly attended to, a large supply of Davy lamps is provided, a sufficient quantity of timber for props, and good machinery.
>
> In ninety-nine cases out of a hundred, accidents are the result of carelessness on the part of the men themselves; who, to save a little trouble, will risk their lives. They are from infancy inured to the dangers of a miner's life, and nothing will ever make them more cautious. There is no protection over the heads of people ascending the pits; I have never known an accident occur in such circumstances. There is no particular age at which children are first put to work. A great deal depends on the parents, and the facilities they have for getting work for them. A man with a large family is anxious to render the boys productive at as early an age as he can, and, if employed himself at such work as will admit of his taking his child or children to assist him, he will frequently do so at from 8 to 10 years of age.
>
> The work at which children are generally employed is in the mines in assisting their parents or others in filling coal or ironstone into the trams, attending air-doors, and in driving horses.
>
> At all iron manufactories there is a good deal of light work, for which children are better adapted than grown up persons, being much more active; and they perform the work with less inconvenience to themselves than would be the case with adults.
>
> Although there are many employed under 12 years of age, I am of the opinion

> they would be equally suitable at this age. I am not aware of any machinery which could be applied to supersede the work of children.
>
> With reference to a limitation of the age at which children are employed in mines – physically I would say there is no advantage arising from it; for, in my experience of the manner children are used throughout the whole of the Monmouthshire and South Wales District, I am of the opinion they would not be benefited by it, as the children are at present hearty and robust, and grow up strong muscular men. To many families such a measure would be a source of hardship, inasmuch as they would be for a time deprived of the wages of the children, which in some cases forms a considerable item in their incomes. In a moral sense I am of the opinion there would be some advantage in it; at present the cupidity or wants of the parents induce them to set their children to work at as early an age as possible, without reference to their education or religious instruction; but if they were interdicted from allowing them to work until after a certain age, their convenience, if no better motive influenced them, would be to induce them to send them to public school if there was one within reach.
>
> I am of the opinion that 10 would be quite late enough to commence work; for, unless children who are doomed to earn their bread by their labour begin about that age, they do not take to it as well at a later period.

Appended to Mr Wood's statement is the following list of accidents at Abersychan Ironworks. There was no timescale attached to these.

| | |
|---|---|
| Hurt but recovered: | |
| Fire-damp | 7 |
| Falls of coal, etc. | 18 |
| Fell into a pit | 2 |
| Hurt by coal tram | 1 |
| Total | 28 |

| | |
|---|---|
| Those killed: | |
| Fire-damp | 1 |
| Falls of coal, etc. | 4 |
| Drowned | 2 |
| Got entangled in a chain | 1 |
| Chain breaking | 2 |
| Total | 10 |

The British Iron Company at Abersychan, near Pontypool, employed at that time some 1,300 people, and produced around 400 tons of iron per week from four furnaces. The youngest children employed were: one boy five years old, two boys seven years, eight boys of eight years, and eleven boys of eleven years.

## Parish of Trevethin Housing, as of May 1840

Some very interesting statistical information with respect to Pontypool and the parish of Trevethin was given by a Mr G. S. Kendrick, Manager of the Varteg works:

> The total population of the Parish of Trevethin is estimated at present at fifteen thousand and I estimate that one third of the gross population, or five thousand persons, are employed in the different works, and one fifth part of the persons so employed, or one thousand, are under eighteen years of age.

Mr Kendrick (a gentleman most highly respected and valued in the neighbourhood) states the general population of the Parish of Trevethin in March eighteen forty, to have been seventeen thousand one hundred and ninety six, the number of houses two thousand nine hundred and eight, giving six persons per house. The number of sleeping rooms is five thousand four hundred and ninety six, giving three and one eighth persons to a sleeping room. The number of children under three years old was one thousand one hundred and eighty four, children from three to twelve years (being the age for education) three thousand five hundred and forty seven, or one fifth of the population. Number going to day school; one thousand and twenty two, average attendance at the Sunday school two thousand two hundred and ten, proportion of houses having bibles, eighty percent, and persons who say they do not go to any place of worship, two thousand one hundred and eighty one. These details were collected a few months after the Chartist riots, since when the extensive ironworks at the Varteg have been suspended (to the great loss of the neighbourhood) and there are, in different parts of the parish, upwards of two hundred houses untenanted.

## Hours of Work

The children and young persons generally work the same hours as adults; seldom less, but frequently more, particularly in the collieries and mine works, where they are almost invariably employed in bringing out the coal and mine worked by the men. They mostly commence at about six o'clock in the morning, and leave off at six or seven in the evening. They very frequently work two or three hours overtime, which they are induced to do from some accidental circumstances interfering during the day with the dispatch of their labours, or when on some occasions a greater quantity of work is required to be done.

On these occasions they generally remain at work until the allotted task is done; and at many establishments, when in full work, and when only the day set of hands is employed, the labours of the pits and levels which commenced at six o'clock in the morning are not suspended until seven or eight in the evening.

Night work is not generally practised in the collieries and mine works, but at the iron furnaces and forges the work is continued both night and day; the hands changing at six o'clock in the morning and six o'clock in the evening, and the children work at night as well as day.

The labours of the children and young persons in the works of this neighbourhood are not continuous when severe, very few complain of the severity of their work, but to a casual observer many descriptions of labour in which they are employed would seem to be extremely severe, particularly in the ironworks, where their active exertions in handling the heated masses and bars of iron, as they pass through the different processes of the forge and rolling mills, often throws them into profuse perspiration; but there are periods of cessation as well as periods of exertion, and after working for half an hour, or an hour, in an atmosphere heated to a high and almost scorching temperature by the bright masses of iron passing in quick succession through the rollers at which they work, the heat being over and the rude mass of metal reduced to a regular shaped bar or rail, the period of cessation arrives, and they are at rest and frequently play for fifteen or twenty minutes, while the furnace again performs its office, heating other piles or masses of metal to undergo the same process; these periods of cessation occur in the forges and mills at least seven or eight times in twelve hours.

The children work the same hours as adults, but there are frequent short intervals during which they rest and take meals. In the coal and mine works they never work by night, but at the fire-work they take their turn the same as the adults.

The furnaces are never suspended at this works on Sundays; at a very few works, the working of the furnaces is suspended on Sundays for about eight hours; when in good order they do not suffer much injury from such stoppage, but if out of order, serious inconvenience is often the result; in all places where Sunday labour is suspended furnaces out of order are kept in blast.

From my experience of the habits of the workmen, I am of the opinion that, so far from their deriving any comfort or moral improvement from suspension of Sunday labour, the contrary in the majority of cases would be the result; instead of attending a place of worship, a beer-shop, The Great Curse of The Country, would be their resort; the government would be conferring a greater boon on the lower classes, and improving their morals, to a much greater extent, were they to close the Kidley-Winks [Beer shops] than to suspend the furnaces.

Such was the thinking of the employers – and their managers etc. – in those early days. Much is made of the benefits of religion, without regard to (perhaps?) the more important benefits of good food and education, and few gave any thought to the welfare of the employee; far more was given to the welfare of the furnaces!

However, not all owners were quite so uncaring of their employees. In Pontypool, many workers were fortunate in having Mr Hanbury Leigh as an employer. He was one who, within the limitations of the time, did care, and showed it by building a school for the children of Twmpath and Trevethin (and we should remember here that in those days, the employment of children was normal, and indeed for some families necessary for survival).

It was not only at the north end of the valley that children were employed in heavy industry. Below are a few examples from Pontrhydirin Tinworks, in the Parish of Panteague, Monmouthshire, belonging to Messrs Conway Brothers, and employing about 50 people on 27 April 1840.

In the plating works, long thin sheets (of iron) were folded and then dipped in molten tin in the tin pot (see below). They were then sheared (guillotined) to remove the fold at each side before being separated into sheets or plates. The sawdust and lime prevented the sheets from sticking together when put in boxes for transportation. These operations are referred to in the statement below.

**Samuel Conway, aged 45, Shearer**

I have eight children, four are working in these tin-works, one boy and three girls. The youngest is with me, she is about eight years and a half old; the boy is 13 years old, he is employed in raising plates out of the Tin-Pot.

The oldest girl is 17 years old, she is employed in rubbing the plates in sawdust and lime; the other girl is 15 years old, she works with me in opening the tin-plates after I shear them, and this little one helps her; they both work by my side; the little one has only been at work four months; I do not mean to keep her at work; I have two working, and I mean to send them to school every other week; I have done this with all my children when they come to work. I do not keep them long at it at a time, but change them every other week, and they get good schooling in this way; they go to a day school, and I pay 4d. per week for them for reading, and 6d. per week for writing, all ages the same. The Master is a tolerably good one, but another school is very much wanted in the neighbourhood, and I have been speaking to Mr Conway about it. The children get, for opening plates, from 3s. 4d. to 5s. 10d. per week; I get for shearing about 25s., per week, and I receive the children's money; they all work from six to six, but go home to their meals, as they live near the works; and as they work by the box, they can come and go as they like, but they cannot do so in the mill, as they must attend the fires, and the metal, and other things. I worked when I was eight years old in a copper-works rolling mill quite as hard as any of the children do here; it did not hurt me, and I do not think it hurts any child to put them to light work about eight years old.

The dipping the plates in lime is not a healthy work; the lime dust gets in their throats and gives them a short cough. They have not done that work here more than two years; it is done by big girls 15 or 16 years old; it agrees with some very well, others complain but get used to it.

**Phoebe Conway, aged eight years and six months**

I work with my father and sister; I open the plates from the shears; my father shears them; I help my sister; I have not been long at the works – not many months; my work is not hard; I would rather go to work than school; I have been to school a little, and go of a Sunday to the chapel school; I can't read, I am spelling; I am going to school again next week.

**Thomas Evans, aged 12**

I am dusting plates with sawdust; I work 11 hours; I have an hour for dinner; about a month ago I worked at the mill; behind the rolls, catching, it was harder work than dusting , it was very hot sometimes behind the rolls; I was at school for three years, and I attend the Sunday school; I can read; I would rather go to school than work; I don't know how much I get; my father receives my money.

**John Absolem, aged 11 years**

I am a lister, I have been working four years, but I have only been listing for four months; I work in the tin-house; we all work twelve hours; I go home to dinner, but am not allowed an hour, sometimes not ten minutes; they must keep on at the work while the tin and grease is hot and in temper; I do not always work twelve hours; I do not get very tired; I am never ill; I do not recollect that I have lost a day by being sick; before I went to list I was putting down plates in the pickle; that was not hard work, but sometimes the pickle burned my fingers; I used hand-leathers, which saved my hands; I did not hurt them much; I was in school three years, and I go to the Sunday school now, and I can read; I would rather go to school than work; I do not know how much money I get, my father has it; he works in the same works; he is a scaler at the furnace.

When this child talks of 'the pickle burning his fingers' we must remember that pickle is acid, usually dilute sulphuric, and usually hot. This report by Mr W. R. Jones continues:

The collieries and mines are mostly well ventilated; when any parts of them are otherwise, it is the result only of casual circumstances, and free ventilation is speedily restored and maintained, and it is seldom that the juvenile part of the work-people are set to work where any impurity of air exists, and, as intimated under the preceding head of this report I do not find that their exclusion from daylight and open air, when employed in mines, and their exposure to subterraneous damp has any effect upon their general health.

In one report from an old book, there was a photograph of an old miner, bowler-hatted, with the caption: 'A veteran of many an explosion'. Neither statement needs further comment from the author!

Fig 138: Ironstone outcrop. The red staining is due to iron oxide, and shows the vein was fairly rich in iron. This particular outcrop has been exposed by surface water washout, but nearby are several ponds which were used for scouring or hushing the mineral lodes. This water is quite clean enough, though, for weed to thrive in it. Taken on the mountain near the Upper Brickworks area of Blaenavon.

Fig 139: General view of what are known as The Scourings. These were ironstone workings, at Upper Race, Pontypool, Gwent, and were exploited by the method of mining known as hushing. This consisted of allowing a large body of stored water to escape from a holding pond to expose whatever minerals were being sought.

Fig 140: A piece of ironstone from Blaenavon, Gwent. This ore, $Fe_2O_3$, was also known as kidney ore from its shape and dark red colour.

Fig 141: The Pontypool Road–Lower Mill–Pontnewynydd tramway showing the entrance to the tunnel from the line of the railway alongside the Afon-Llwyd river.

Fig 142: The view inside the tunnel looking back toward the entrance. The far end has been blocked off for many years, and originally exited near the site of the Park Cinema, thence to the Town Forge rail sidings, before passing – and serving – the Town and Central forges, as well as the Osborne Forge, before connecting with the main Newport–Blaenavon line at Trevethin Junction, near Pontnewynydd.

Fig 143: The two tunnels under the Newport–Brecon canal at Pontymoile; they led to Lower Mills sheetworks. The tunnel on the left was for the tramway, which was of the standard gauge of 4ft 8½in, from Pontypool Road to Pontymoile ironworks (plus the various forges between) and to Pontnewynydd. The tunnel on the right was for the watercourse, or leat, which conveyed water from a weir on the Afon-Llwyd river to a waterwheel in Lower Mills works. The tramway continued behind the camera to a weighbridge at Pontymoile, and a nearer branch line of some 70 yards; this branch led to a crane and wharf on the canal basin, which is 60 yards or so to the right of the camera. The weir is still in place today, as is the gate to control the water supply.

Fig 144: The Lower Mill tramway sluices on the River Afon-Llwyd near the new Pontypool Leisure Centre. The left cast iron sluice controlled the water for the waterwheel at Lower Mills. The two racks were to lift the sluice door at the bottom of the cast iron sluice-gate. They engaged with cogs on a hand wheel. This was bolted to the top of the two cast iron frames near them.

Fig 145: This shows both sluices. The one on the right controlled the water level in the small brick dam across the river. This brick dam has long been demolished, but is still lying on the river bed. The water wheel at Lower Mills was 20ft diameter and 10ft wide.

Fig 146: A stand of mills used in the pack-mill manual system of rolling sheets. These are at the Pontnewynyd Sheet and Galvanizing Works, near Pontypool.

There were two engines, each developing 1,500 hp at 32 revolutions per minute, and each driving six rolling mills. The large castellated wheel on top of each mill was for adjustment of the rollers, using the levers shown. One roll can be seen suspended on the overhead crane.

The small mill at the near end was a temporary addition. The side near the camera was the roller's side, the roller being the man in charge of the mill. Behind him and to the right were the furnaces; these were coal fired. Note the small set of wheels near the mill. These supported the rolled-out sheet as the roller fed it into the rolls.

A pack started off with two steel bars, approximately 8in wide, 2ft 6in long and ⁵/₈in thick. After heating, these were fed into the rolls one after the other. They were then rolled into two plates about 2ft 6in by 3ft. They were then reheated before being rolled together. This procedure was carried out a number of times until there were 6 or 8 sheets in the pack. These were then sheared to size and parted into separate sheets, after which they were pickled in acid, galvanized and corrugated.

It would be difficult to portray conditions in the forge as we knew it, especially in the summer, as these sheets were 10ft or so long, and extremely hot when brought to the roller. Mounted on the overhead crane girders were several red lights. In the event of steam pressure in the boilers falling through heavy loading, these lights would illuminate, at which rolling stopped. I can remember seeing, in high summer, all the rollers feeding bars into the mills together, causing a steam shortage, and when the lights came on, there was a big cheer from the mill hands.

The author's father, when working at Lower Mill Works, Pontymoile, in the 1920s, fell on a hot pack, and carried the scars to his grave. It was difficult to extricate oneself from this position, as putting a hand on the sheet to push oneself off resulted in the skin being burned off, acting as a lubricant. This was, apparently, quite a common accident in this trade.

Fig 147: One of the engines built by Hick Hargreaves, of Bolton, in 1910. The engine flywheel weighed, with the crankshaft, around 160 tons.

Fig 148: A pair of crocodile shears, in this case with both steel cutting edges missing.These were driven from a shaft under the steel floor, belt-driven from an electric motor mounted on one of the girders. One of these motors can be seen on the lattice girder over the ladder just to the left of the flywheel in Fig 147.

# 16
# Iron and steel making in Pontypool

Ironworks were established at Pontymoile, on the outskirts of Pontypool, in Gwent, around 1425 by David and Ieuan Grant, using charcoal for smelting, and steel and iron working has, in one form or another, been continued until recently, a period of over five hundred years. Its demise was partly attributable to the rise of strip mills, which heralded the end of the pack-mill method of tin-plate working. Archdeacon William Coxe, who was entertained by Mr Capel Hanbury Leigh at Pontypool Park House, was also given access to documents and stated in his *Tour Of Monmouthshire,* 1801, that the earliest conveyance deeds of an estate at Pontypool, dated 1565, were to a Mr Capel Hanbury, of Kidderminster, who was the founder of the ironworks. His son, Richard, was to enlarge the works considerably later; and in due course, the property was inherited by Richard's son, Capel Hanbury, who died in about 1704, aged 79. Capel Hanbury's son, Major John Hanbury, MP, who was born in 1664, made many advancements and increased the output of the ironworks by means of improvements in the machinery. He is credited with the invention of rollers for rolling plates; he also introduced the tinning process into Britain. He died in about 1734.

But quoting from Mr Chris Barber's excellent book *Eastern Valley, The Story of Torfaen* the invention of the rolling mill was credited to one Thomas Cooke, a manager of the Hanbury works, in 1697.

Major John Hanbury's son Capel died in about 1765, Capel's son John died in 1784, and John's elder son John died a bachelor in 1765. His second son was Capel Hanbury, who in 1797 assumed the name of Leigh on inheriting the property of Lord Leigh.

In about 1665 Major John Hanbury started the manufacture of tinplates at Pontypool, with the assistance of Andrew Yarranton, who had been to Saxony acting as the agent of a syndicate of wealthy Staffordshire men, to ascertain the method of manufacture, and returned to Britain with a number of workers skilled in the making of tinplates.

Before 1713, when Abraham Darby first devised the means of smelting with coked pit coal, iron smelting was done using charcoal, which process incurred a lot of expense and extra labour, as well as denuding the local forests and hills of trees.

With regard to the use of charcoal for smelting or heating: in general, and depending on the water content, wood has only about one-fifth to one-quarter the calorific value of coal, and therefore needed considerably more per ton of iron smelted.

The Town Forge in Pontypool was situated in Park Road, and within twenty yards of The Cross, which is considered to be the centre of the town, while the Park Forge was erected on a portion of the park, almost opposite to the Rock Hill.

It seems that both these forges were quite near the Park House, the home of Mr C. H. Leigh who, on account of the nuisance caused by smoke and noise, had them dismantled. There is little if any trace of either works to be seen today.

If one leaves the Pontypool Leisure Centre car park by the steps at the south end into what remains of Trosnant, by approaching the river bank through the wall, one can still see some of the tunnels from the Park Forge or possibly the wire-works, under the park on the opposite bank of the river. Also close by is the lower end of the brook from the Glyn Ponds. This brook flowed down the valley past the Mynydd Maen (Stony Mountain) Colliery, where at one time it also drove a large waterwheel before crossing the Crumlin Road near the junction of Race Road, past the council destructor, where much of Pontypool's waste was burned. There was also a steam-driven mortar mill, the engine of which I was the last person to retube, at Albion Road, before continuing past the Pontypool gasworks at the bottom of Mill Road and Trosnant (Trosnant = Cross Brook) where it passed under Trosnant Road finally to exit to either the canal or the river.

Today, most of the water in this brook appears to be from the old Glyntillery No. 2 mine, some four hundred yards down the line (westward) from the Quarry Level.

The water that was pumped out of the Old Glyn Pits also fed into this brook. It was conveyed from the up-cast shaft of the Old Glyn to a spot near the bottom of the Twmpath Road via an underground watercourse, and it is reliably understood that it exits near the tunnel from near the Drill Hall, Pontnewynydd, close to the sawmill on Albion Road.

A report states that both shafts at the Old Glyn have filled with water up to the level of the watercourse; if so, then this water will probably still flow through this watercourse to the original outlet, even if it has partly collapsed.

I believe the Glyn Ponds were originally installed to maintain the water level of the Monmouthshire and Brecon Canal, and just at the exit steps of the Leisure Centre car park, about half a mile away, and where the brook exits the tunnel beneath the road in Trosnant, the cast iron sluice-gate to the leat can still be seen. Also worth looking at is the high standard of the masonry, even though it was nearly out of public view (see page 147).

If one leaves the Leisure Centre car park by the north (Pontypool) end, on the old tram-road, one can also still see the cast iron weir and sluices on the Afon-Llwyd river (Afon-Llwyd = Grey River) which supplied water for the waterwheel at Lower Mill ironworks in the years before steam was employed there. This leat runs parallel to the river, and the tram-road, for a short way, then diverges just after the Pontypool/Usk/Newport road crosses the river, and near the confluence with the Dare Brook. It then runs almost parallel again with the tram-road from Lower Mills, under the right-hand (western) canal tunnel, past Lower Mill Row, where my father was born, then into Lower Mills, where both my father and grandfather worked.

According to an inventory of 27 July 1852, the water wheel at Lower Mill was twenty feet in diameter by ten feet wide, and appears to have been coupled to a pair of black-plate (i.e. tinplate) rolls through gearing. It also drove a pair of doubling shears, two bar cropping shears and a shearing machine, with gearing and chain in good order. Shears cut sheet iron to size.

This tram-road, which I believe was of the standard gauge of 4ft 8½in, came up from Lower Mill, under the canal via the left – eastern – of two tunnels. It served Pontymoile ironworks and Park Forge, Pontymoile gasworks, which was sited where Messrs Baileys' coal yard was, then ran under the tunnel near the Italian Gardens opposite Pontypool Town Hall, and which is still partly open, to serve Town Forge and Osborne Forge. Near the Lower Mills tunnel end, there was also a short rail branch shown now by the route of the lane, back to the canal basin, where old maps show there was a crane, and indeed this crane can also be found in the 1852 inventory of Lower Mill works, where it states: 1 large crane, on the canal wharf, near the junction. Chain made of three-quarter inch bolts [*sic*]. Gearing, etc in good condition.

The tracks of all these railways are still traceable today. There are three similar tunnels under the canal a short distance away. These are smaller in diameter than the rail/water tunnels, and were included in the canal work probably as relief tunnels, which would, in the unlikely event of the aqueduct bridge over the Afon-Llwyd becoming blocked, allow the flood water to bypass it.

There was also a tram-road from Blaendare which travelled down through Cwmynyscoy, crossing what was recently St Matthew's Road, then over the canal locks. These were on the Pontnewynydd branch of the Monmouthshire and Brecon Canal it then ran into the Pontypool Foundry on Rockhill Road. The rails from this tram-road were visible until recent years crossing Victoria Road, and are probably still in position under the tarmac road surface. This tram-road started near the Aberdare No. 1 ironstone level on the Upper Race.

These burgeoning industries, and in particular coke ovens, as those at Blaendare, laid the foundations for the many small coal levels in the area, particularly around Blaendare and Upper Race, with deep mines, like the Balance Pit in the late 1700s. These, as demand for coal increased rapidly, were soon superseded in 1845 by the Old Glyn Pits, both sunk by the Hanbury family, which, at 190 yards deep, was both one of the deepest and also, with the Cwmlickey pits – which also wound coal from 98-yard shafts using a beam engine – on the Upper Race, one of the first pits in South Wales to use the power of steam for winding coal and pumping water.

## The Fox family connections with Pontypool

In 1820, gasworks equipment was made at the Neath Abbey Ironworks of the Fox family for a Mr Watkin George, Junior, then of Pontypool, but late of Dowlais, and I believe that this was for the gasworks at Pontymoile. This was situated behind the Pontypool Foundry on Rockhill Road (in 1851, Neath Abbey Ironworks also made gas plant for the Usk Gas Company). This would place it next to the Pontymoile ironworks, and also next to the Pontypool Road/Lower Mills/Town Forge/Pontnewynydd railway.

The reader will probably have noticed that the beam engine at Glyn Pits was also made at Neath Abbey Foundry by the Fox family. The Fox family were Cornish, and in 1791 a foundry was set up at Perran Wharf in the Fal estuary.

The partners of the enterprise were:

George Croker Fox
Robert Were Fox
Thomas Were Fox
Mary Fox
George Fox
Thomas Fox
Edward Fox
Peter Price
Samuel Tregelles
Thomas Wilson of Truro, a Copper Smelter
John Gould of Truro, Doctor of Physic
William Wood of Swansea, an Ironmaster

Most of the partners were related and most came from Falmouth itself. In July 1792 the same partners took over the lease of the Neath Abbey Ironworks. The family were also involved as shipping agents, bankers, and in general merchandise, etc.

Since the Fox family were also members of the Society of Friends (Quakers) it seems likely that this would have ensured preferential treatment from other members of the same sect, especially in pricing and costs. This observation applied equally to the Darby family of Coalbrookdale, who were ironmasters, and, of course, like the Fox family, members of the

Quaker faith, and from where the Foxes purchased much of their pig iron, probably at a very favourable rate, evidence suggesting that this was indeed the case.

The output of the Neath Abbey Ironworks was both large and varied. Between about 1805 and about 1850, there were close to 700 stationary engines alone; these included both beam and horizontal engines, winding engines, blowing engines for blast furnaces, etc. These engines ranged from tiny engines with a $3^1/_2$in cylinder bore right up to a mighty beam blowing engine with 122in air cylinder, and a 24in high pressure engine for the Victoria Ironworks, at Ebbw Vale.

Among the others were:

A 40in blowing engine and a 34in beam engine for the British Ironworks.
A $52^1/_2$in blowing engine for Mr C. H. Leigh, Pontypool.
An 11in engine for a Mr Strick (now the works of Messrs Fred Lloyds, in the Polo Grounds, Pontypool).
A 40in engine and two $52^1/_2$in blowing engines for Blaenavon Ironworks.
Another 30in by 6ft beam pumping and winding engine for Mr C. H. Leigh, Pontypool in 1845. (This is the beam engine still in place at Glyn Pits).
A 36in by 6ft beam engine for the Pontypool Ironworks.
A 16in engine for Mr C. H. Leigh, Pontypool.
A 45in engine with a 20ft flywheel for Abersychan Ironworks.
A $52^1/_2$in blowing engine for Messrs. Leigh and George, at Pontypool Ironworks.
A 40in engine for Blaenavon Ironworks.
A 34in condensing beam engine.
A $52^1/_2$in beam engine for Mr C. H. Leigh, Pontypool.

There was also a beam engine in the Forest of Dean with a 70in cylinder and a 12ft stroke. All this as well as gas plant, ships, both steam and sail, locomotives, dredgers and canal boats, etc.

It might be as well to explain that blowing engines were used to supply blast furnaces with air under pressure, usually about two or three pounds per square inch. In later years, possibly partly due to the greater use of electric motors and better high-speed fans, these were used in preference to the very inefficient steam reciprocating engines for the supply of blast air. This equipment was not only supplied to British companies. It was also exported to Spain, Germany, South America, Mexico, etc.

## The Quakers, or more properly, The Society of Friends

The *Journal* of George Fox, founder of the Society of Friends, states:

> Justice Bennet of Derby was the first that called us Quakers; because I bid them tremble at the word of the Lord. This was in the year 1650.

The Quaker sect was founded by George Fox in the early 1660s and was organized as a distinct group in 1668, when they began to engage in missionary work. The Quakers came to Pontypool following the split in the Puritan Movement, and one of the earliest members was Mr Richard Hanbury, cousin of Capel Hanbury of Pontypool, who, in 1669, held services in his own home until the Quaker meeting house was built at Pontymoile, opposite Maesderwen House. A new meeting house was built in Trosnant to replace the one at Pontymoile, but was abandoned in about 1820. There were Quaker burial grounds at both Pontymoile and Trosnant. Again quoting Mr Chris Barber's excellent book *The Story of Torfaen*, he tells us that Mr Richard Hanbury, who died in 1695, was buried in the Pontymoile Quaker burial ground. It is believed that this particular burial ground was to the right of the Three Legged Bridge at the Pontypool end of Great Western Terrace, and behind a small scrapyard owned then by Mr Harris, and was still there in c.1939.

There was another Quaker enterprise in Cwmafon, at the old Westlake's brewery, although your author is not certain what this was. As employers, they had a name for being both fair and just, and it seems that this was a universal opinion, because they even impressed no less a personage than the dour Puritan, Oliver Cromwell, Lord Protector of England.

Quoting again from George Fox's *Journal*:

> He [Oliver Cromwell] said: 'I see there is a people risen, that I cannot win either with gifts, honours, offices or places; but all other sects and people I can.'

A Mr Watkin George, Junior, was engaged by Mr C. H. Leigh in about 1807 to reconstruct and improve the old furnaces at Blaendare, and it is probably from about this time that the modern industrial history of Pontypool can be dated.

Records show that the Osborne Forge predated the Town Forge, but by how long is not known, nor is there a date for the erection of the Osborne Forge; apparently, it is from the name of this forge that the brand of iron known as Osborne or the earlier name Osmonde derives its name, and also Osborne Road, previously Manchester Road. It seems that the original name for this type of iron was Asleom, of which the later names Osmonde and Osborne were a corrupt version. It may not be generally known that the Osborne Forge – near Pontypool Drill Hall – also had two waterwheels. These were 12ft diameter and 3ft 10in wide.

At the Pontnewynydd end of the Osborne Forge was the reservoir we knew as Forge Pond, originally to feed the two waterwheels, and when it was built, this pond contained 12,300 cubic yards of water when even with the overflow. In later years it was badly silted and contained far less than this. However, by this time the Osborne Forge was no longer viable, being converted into a bucket factory before its final closure.

This overflow was a large cast iron affair on the side of the mineral railway. It was about 6ft long and 18in wide and 6ft deep, and it led under the railway to the nearby river. The open top, like many other things in those far-off days, would not now be permitted, and it was really very dangerous.

This pond was fed from a sluice-gate (now demolished) at a higher point on the river near Freeholdland; it crossed under Mill Road near the United Friends public house, passed behind the Pavilion cinema, crossed the brook near Messrs Chaloner's electrical works, then ran parallel to Osborne Road for a short distance, under the entrance to Peake's yard, then into the pond. As children, we all had many wet feet playing near the leat, which we knew as The Dyke, and pond. There have also been some fatalities in that pond.

Osmonde iron was apparently of very high quality, and was used by the governments of both Elizabeth I and James I. There is good reason to assume that wireworks were established in Pontypool at about the same time as ironworks.

It is known that the first Richard Hanbury had connections with Tintern and also of his successful management of the works there. Also of his wide experience of the drawing of wire and iron manufacture, largely gained at Tintern.

The name Tintern is significant. Terne metal is iron plate coated with a mixture of lead and tin (often Forest of Dean iron and Cornish tin), a coating which gave several advantages, for example as a lubricant in the drawing of wire and also in the preservation of iron plates, i.e. as a base in painting. Lead will not bond to iron plate, but tin will, so that a mixture of lead and 20% tin, was used.

The wireworks were at Pontymoile, and were removed in the early 1800s, and the local production of Osmond iron, which was ideal for making wire, was probably one factor in the siting of the wireworks. Both works were of course linked by the railway.

Pontypool was also justly famous for the making of Pontypool Japan ware. The Japan industry was introduced into Pontypool in about 1661 by one Thomas Allgood, who was from Northampton. There is a strong possibility that he was induced to come to Pontypool by Major John Hanbury.

It seems that Thomas Allgood had already distinguished himself by making some remarkable discoveries. It was he who first devised the means of making copperas, or iron sulphate, which was used in tinworks, and also in oil colours. He was also famous for having invented a method of extracting oil from cannel coal. But it is his talent for lacquering iron for which he is remembered. He was reported to have had a good chemical knowledge and it was this that enabled him to develop the Japanning process.

Japanning was carried out in West Place, just off Crane Street, to which there were two entrances. When I was young, it was still known as Old Japan. One of the Allgood family is buried in the Baptist church at Penygarn, with the headstone erected in front of the church.

# 17
# Full Circle

It is now nearly 580 years since ironmaking was started at Pontypool, in Gwent, by Ieuan and David Grant. During the later years of the last century or so, there were many fine – and famous – steel and iron works to be found here.

The steelworks end of Panteg works has now gone, with the sheet works also being demolished. The works, which started rolling sheets in 1895, had, in 1932, nine melting furnaces with a rolling mill, sixteen sheet mills, and one plate mill. In his book *Rolls and Rolling Mills*, Mr Harold Cookson tells us that the first sheet steel in Great Britain was rolled at Thompson and Hatton's Bradley Tinplate Works in 1876 from a steel bloom made by a Mr Isaac Butler at Panteg Steelworks. The same source states that in 1889 Baldwins commenced rolling sheets for galvanizing at their Lower Mills works, Pontypool. Iron sheets were being rolled there before 1870, and the works were apparently in existence in 1818, according to an inscription on the keystone of an arch. It is not clear if the arch referred to here is one of the two arches – railway and leat – under the canal; if it is the inscription cannot be seen.

All the collieries have also gone, again with the exception of a few private mines, (all small levels or slopes), leaving little or no heavy industry in the valley at all. Now, with the demise of the local rail service valley branches, it is unlikely in the extreme that it will ever return, as road transport, not to mention the roads themselves, would be quite unable to handle the volume of coal and steel that would be required for economical working.

However, there are some advantages accruing from this, such as clean rivers and unpolluted air and fewer serious accidents, particularly in the mining industry, although it must be said that many of these were attributable to ignorance and parsimony on the part of the mine owners. For example, in the early years of this century in mining, candles were used in some mines in preference to safety lamps solely on the basis of cost. In this context, it should also be said that conservatism on the part of some of the miners also helped to delay the introduction of the safety lamp, because some of them preferred the small amount of extra light given by a candle, since the early safety lamps were not very efficient, with the flame in the very first Davy lamps protected by gauze. It was also many years before the role of coal dust in enhancing firedamp explosions in mining was properly recognised, and safety measures, such as the spreading of stone dust to break the chain of combustion in fine coal dust, etc. were instituted.

Apart from the noise of the heavy traffic on the new Crumlin to Pontypool road, the route of the mineral railway, or Park Estate Line, as it used to be known, in the Glyn Valley is now once more a pleasant place to walk. But even the traffic noise does not intrude unduly, since it is masked to a great extent by the immense spoil tips, which are a comparatively recent addition. There is still much water flowing from the entrance to Glyntillery No. 2 mine, but now it is clear and quite drinkable, and, as proof of this, there is much plant life in it.

As one approaches the Old Glyn, there is a strong smell of pine from the forest on one's right. The scars left by coal mining and quarrying for stone at the Quarry Level and the Glyn Quarry above it are, unless one knows just where to look, obliterated, as are the screens for both the Quarry Level and the Old Glyn Pits. The Blaendare wagon repair shed, where I started my working life, has long since gone, together with its rail sidings, although there are traces still left, such as wagon bolts left inserted in the wall where the sidings were. The sidings have also disappeared from the Old Glyn Pit, leaving just the indentations of the railway sleepers underfoot, and the occasional rotting sleeper with its shoes still attached.

There was no heapstead at the Old Glyn Pit. Coal was delivered to bank at ground level, then hauled by a steam haulage-engine up a short incline to the top of the screens which were a hundred yards or so away to the west, and built out over a siding alongside the main mineral railway, the Park Estate Line.

The screens are now completely missing, but part of the incline is still in position, overgrown now with larch trees etc. Behind these screens, and running along the side of the mountain, was the water supply from the Quarry Level to the Old Glyn reservoir. It too, as shown on Fig 112, is still, for the most part, easily traceable. The redundant water supply brought from the Balance Pit to the east of the Old Glyn, however, has been almost completely obliterated by forest and bulldozer, and is now traceable only with great difficulty, as are two of the three holding reservoirs for the Balance Pit. These holding reservoirs were quite substantial, with the Upper Reservoir holding 7,720 cubic yards, the Middle Reservoir holding 1,420 cubic yards, and the Lower Reservoir 1,800 cubic yards, a total of almost 11,000 cubic yards of water.

The Old Glyn Pits, sunk by the Hanbury family in the early 1840s, are now, after some 160 years or so, and the recorded deaths of 69 miners, and at least 9 horses, once again back in the hands of the same family, and, because of a few dedicated people, are now, in 2005, belatedly receiving some attention, with both buildings scaffolded for support, and the stonework on the top of the two buildings being reset as a start.

How far this restoration will go is not known yet, but we are hoping for a much more ambitious program in later years. Together with Torfaen Borough Council, Cadw, etc. a steering committee has now been formed to coordinate work on this project. It is to be hoped that a determined effort will now be made to preserve and protect from further attack by vandals, thieves, weather, etc. what little of our industrial heritage we have left.

Also hopefully we – that is, my colleague, Mr Clive Davies and I – are trying to draw attention to the remains of another small piece of our industrial heritage; the water balance wheel at the British, Talywaun, of which there are very few examples remaining in Britain. Partly because of its remoteness, many of the parts of this water balance have been stolen for scrap; this includes one of the A-frames (the main end frames) weighing around a ton, the two bearings, the water-filling mechanism, and the brake mechanism. We also have some knowledge of a large waterwheel on the British which has been buried, but of that another time.

I have included an engineering drawing (Fig 126) of the water balance wheel at the British, Talywaun, dated c.1850. This drawing is not of the balance wheel remains shown on the photographs, i.e. the lower wheel; it is of the wheel; on the top balance pit that is, nearer to the site of Cwm Bergwm colliery, the boiler stack of which is still there. While this water balance has nothing to do with Blaendare, I have included it to illustrate how such wheels worked.

The maker of the vertical winding engine at the Old Glyn has long been a mystery, with many people attributing it to the Neath Abbey Ironworks. However, there is no mention of this in the list of engines made by them, a list which numbers close to 700 alone. Recently, my colleague, Mr Clive Davies, pointed out to me that the Old Glyn Pits were leased to the Ebbw Vale Steel Iron and Coal Company in 1855, and one of the four men involved in this enterprise was a Mr Darby, of the same family that were famous for the Coalbrookdale ironworks. They too were of the Quaker faith.

Might this engine have been built at Coalbrookdale Ironworks? It is a distinct possibility, as the dates correspond quite closely, and there are engines with similar characteristics at the Ironbridge Gorge Museum Trust, Blists Hill site, in Shropshire.

As mentioned earlier, both winding and pumping at the Old Glyn was carried out by a beam engine, which by the standard of those days was not a very large one. As soon as the Ebbw Vale Company leased the Glyn in 1855, where there were now proven coal reserves, then the decision would have been taken to increase production. This would have required a much larger and purpose-built winding engine, so what more natural than that a factory linked to one of the four leaseholders – in this case the renowned and powerful Darby family of Shropshire – should build it? They were certainly capable of doing so.

But there were others too, who were quite capable engineers, and who were much nearer to the Glyn than Coalbrookdale, for example Abersychan Ironworks, which commenced operations in 1860 or so, and just about the time the vertical winding engine was installed at the Old Glyn Pits. It is on record that the Abersychan/British Ironworks built beam engines. Might they also have built the vertical winder at the Old Glyn? Or perhaps the Nevilles of Llanelli built it? Or the Prices? To name just two among hundreds of south and south-west Wales foundries. There is a beam engine at Treforest, which I believe is now in the College of Higher Education there. This was built at the Varteg ironworks.

We shall probably never know the origin of the Glyn Pits vertical winding engine. I was told some time ago that many leather-bound files – actually several car-loads – were removed from the British Ironworks site by a person in a car. I can neither verify this information nor vouch for it. Yet, for such a large enterprise as the Abersychan/British Ironworks was, and it was very large, such files must have existed in recent years, and probably still do. One day, and if they have not been split up, some lucky person may find them.

# INDEX